notia only

Ceterus Paribus Laws

Edited by

John Earman
University of Pittsburgh,
Pittsburgh,
USA

Clark Glymour
Carnegie Mellon University,
Pittsburgh,
USA

and

Sandra Mitchell
University of Pittsburgh,
Pittsburgh,
USA

Reprinted from *Erkenntnis* Vol. 57:3 (2002)

KLUWER ACADEMIC PUBLISHERS
DORDRECHT / BOSTON / LONDON

A C.I.P. Catalogue record for this book is available from the Library of Congress.

ISBN 1-4020-1020-6

Published by Kluwer Academic Publishers,
P.O. Box 17, 3300 AA Dordrecht, The Netherlands.

Sold and distributed in North, Central and South America
by Kluwer Academic Publishers,
101 Philip Drive, Norwell, MA 02061, U.S.A.

In all other countries, sold and distributed
by Kluwer Academic Publishers,
P.O. Box 17, 3300 AA Dordrecht, The Netherlands.

Printed on acid-free paper

Printed in the Netherlands

Table of Contents

EDITORIAL

The use of *ceteris paribus* clauses in philosophy and in the sciences has a long and fascinating history. Persky (1990) traces the use by economists of *ceteris paribus* clauses in qualifying generalizations as far back as William Petty's *Treatise of Taxes and Contributions* (1662). John Cairnes' *The Character and Logical Method of Political Economy* (1857) is credited with enunciating the idea that the conclusions of economic investigations hold "only in the absence of disturbing causes".[1] His *Leading Principles* (1874) contains the classic example of a *ceteris paribus* law: "The rate of wage, other things being equal, varies inversely with the supply of labour". Carines' ideas were popularized by Alfred Marshall in his *Principles of Economics* (1890) where he argued for a methodology that involved holding disturbing causes "in a pound called *Caeteris Paribus*". It is unclear when the notion of *ceteris paribus* laws made its appearance in the philosophical literature; but in the nineteenth century it is to be found in Mill's *System of Logic* (1843), and in the twentieth century it gained prominence in the Hempel-inspired debates of the 1950's over the role of general laws in historical explanations, albeit under other labels such as *quasi-laws* (Rescher) or *grounded generalizations* (Scriven).

The topic of *ceteris paribus* laws is the focus of a minor but seemingly recession-proof industry in philosophy: hardly a year passes without the appearance of at least a few articles on this topic in major journals. One of the driving forces of this industry stems from a worry about the status of the special (or inexact) sciences. The worry starts from the assumptions that science aims to discover laws of nature, and that laws are necessary for the functioning of a mature science since, for example, scientific prediction and explanation rest on laws. The worry is realized when these assumptions are combined with the observation that the special sciences have not produced – and, perhaps, are incapable of producing – any plausible candidates for laws in even the most minimal version of the standard sense of that term – strictly true universal generalizations possessing wide scope and explanatory power. What can be called the CP defense of the scientific status of the special sciences takes two forms.

The first form denies that there are any relevant differences between the special and the fundamental sciences since it is *ceteris paribus* all the way

Erkenntnis **57**: 277–280, 2002.
© 2002 *Kluwer Academic Publishers. Printed in the Netherlands.*

down – some, or perhaps most, of the basic laws of physics contain (perhaps implicit) *ceteris paribus* clauses. The "CP all the way down" thesis is attacked by Earman, Roberts, and Smith (this issue) who claim that it rests on confusions, misunderstandings, and outright falsehoods. Nancy Cartwright has been read as taking the opposite side of this issue. Her contribution to this issue makes it clear that she has been misread, about which more below. And in "There is No Such Thing as a *Ceteris Paribus* Law", James Woodward (this issue) contends that regular and explicit use of the *ceteris paribus* locution is confined to economics.

The second form of the CP defense drops the "CP all the way down" line (thus, admitting that there are important asymmetries between the fundamental and the special sciences) in favor of contending that *ceteris paribus* laws are scientifically legitimate. Here two issues arise. One issue concerns the content of *ceteris paribus* claims, what possible states of the world, or of its history, they include and exclude. Another concerns how such claims can be tested or confirmed. For if *ceteris paribus* claims have no content, they cannot be laws, and if they have content but we cannot come to know them, or have rational degrees of belief in them, rationally changed by evidence, then they might as well not be laws. Earman, Roberts, and Smith (this issue) express skepticism on both issues. They are joined by Woodward (this issue) who argues that extant attempts to save *ceteris paribus* laws from vacuity are massive failures.

This issue also contains several lines of response to the challenge of the scientific legitimacy of *ceteris paribus* laws, as well as papers and arguments directed more particularly to the notion of "law". Gerhard Schurz (2001, and this issue), while being critical of extant attempts to save *ceteris paribus* laws from vacuity, is sanguine about finding empirical content in *ceteris paribus* laws, provided that they entail statistical normality claims. Marc Lange is equally sanguine, but for different reasons. In "Who's Afraid of *Ceteris Paribus* Laws?" (this issue), Lange argues that, despite their open endedness, *ceteris paribus* generalizations can have a definite meaning if there is sufficient agreement – perhaps tacit – both on canonical examples that fall under the generalization and on the analogies/disanalogies with canonical examples that would supply good reasons for or against counting a factor as "disturbing". For Lange *ceteris paribus* generalizations can serve as laws of nature not because they correspond to regularities but because they function as reliable inference rules. The virtue of such an account, Lange argues, is that it directs our attention to how laws actually function in science and away from philosophers' quest for the characteristics that generalizations must have in order to deserve the honorific of 'law'.

[2]

A third defense of the legitimacy of the special sciences rejects the assumptions on which the *ceteris paribus* industry is based. Woodward is an articulate advocate for this line. In his contribution to this issue, Woodward presents an account of the nature of the special sciences on which their proper functioning does not require the establishment of laws. Rather what the special sciences thrive on are causal generalizations, which typically do not qualify as laws, but which can be subjected to rigorous testing and which serve as the basis for scientific explanations. Sandra Mitchell's "*Ceteris Paribus* – An Inadequate Representation for Biological Contingency" (this issue) challenges the claim that it is the contingency of causal regularities in biology that precludes them from achieving lawful status. She defends weakening our conception of law so that it includes the types of generalizations Woodward takes as necessary for explanation. Articulating different types and degrees of contingency, rather than relegating all non-strict dependence to a *ceteris paribus* clause, permits a more nuanced approach to characterizing the differences between the exact and inexact sciences.

This issue also contains two formal responses to the problems. Wolfgang Spohn argues that a universal generalization is a law not due to its particular content, but because a certain inductive behavior is associated with it, a behavior which may be precisely described within a dynamic theory of belief or acceptance. He continues to explain how this framework may be used for accounting also for *ceteris paribus* conditions and for responding to both the semantic and epistemic challenges posed by Earman, Roberts, and Smith. Clark Glymour takes on both the semantic and epistemic challenges posed by Earman, Roberts, and Smith. He gives "*ceteris paribus*" a subjective interpretation and, insisting that the issue of empirical content turns on whether *ceteris paribus* claims can be reliably learned in the limit, argues that we can reliably learn the truth or falsity of both our own *ceteris paribus* claims and those made by others.

Nancy Cartwright's paper "In Favor of Laws that are not *Ceteris Paribus* after All" (this issue) indicates that she is not an advocate of *ceteris paribus* laws, at least not if such laws cannot be stated in precise and closed form and do not entail strict or statistical regularities. Rather her main concern is to override positivistic scruples that prevent us from seeing that laws are basically about capacities and powers. In "Cartwright on Explanation and Idealization" Elgin and Sober (this issue) address Cartwright's claims in *How the Laws of Physics Lie* (1983) that when laws of conditional form are true, they don't apply to real world situations because the antecedent condition involves an idealization. Elgin and Sober demur on the grounds that a conditional and its contrapositive apply to the same objects, and that if $C \rightarrow L$ fails to apply to real objects because C involves an idealization, then $\neg L \rightarrow \neg C$ applies to real objects. However, they go on to give an argument for a special case of another of Cartwright's claims; namely, true fundamental laws are not explanatory. But they also argue that there are many cases where idealizations are explanatory.

The present collection of articles reveals how the debate over *ceteris paribus* laws is tied to fundamental issues concerning the content and methodology of the sciences. Since there is no realistic hope that these issues will be resolved in the foreseeable future, the *ceteris paribus* industry will, no doubt, continue unabated. But it is our hope that this collection will help to guide the industry towards more productive ways of processing the issues.

NOTE

[1] Schurz (1995; and this issue) has contrasted this *exclusive* sense of *ceteris paribus*, in which disturbing factors are assumed to be absent, with a *comparative* sense, in which disturbing factors are not excluded but are assumed to be held constant.

REFERENCES

Cairnes, J.: 1857, *The Character and Logical Method of Political Economy*, Longman, Brown, Green, Longmans and Roberts, London.

Cairnes, J.: 1874, *Some Leading Principles of Political Economy Newly Expounded*, Harper and Bros, New York.

Lewis, D.: 1973, *Counterfactuals*, Harvard University Press, Cambridge, MA.

Marshall, A.: 1890, *Principles of Economics*, Quotation from the 8th edition, Macmillan, London, 1979.

Mill, J. S.: 1843, *A System of Logic*, J. W. Parker, London.

Mitchell, S. D.: 1997, 'Pragmatic Laws', in L. Darden (ed.), *PSA 1996: Part II, Symposia Papers, Philosophy of Science (special issue)*, S468–S479.

Mitchell, S. D.: 2000, 'Dimensions of Scientific Law', *Philosophy of Science* **67**, 242–265.

Persky, J.: 1990, 'Ceteris Paribus', *Journal of Economic Perspectives* **4**, 187–193.

Petty, W.: 1662, *Treatise of Taxes and Contributions*, Printed for N. Brooke, London.

Schurz, G.: 1995, 'Theories and Their Applications: A Case of Non-Monotonic Reasoning', in W. Herfel et al. (eds), *Theories and Models in Scientific Processes*, Rodopi, Amsterdam, pp. 269–293.

Schurz, G.: 2001, 'Pietroski and Rey on Ceteris Paribus Laws', *British Journal for the Philosophy of Science* **52**, 359–370.

John Earman
Clark Glymour
Sandra Mitchell
University of Pittsburgh
U.S.A.

JOHN EARMAN, JOHN ROBERTS and SHELDON SMITH

CETERIS PARIBUS LOST

ABSTRACT. Many have claimed that ceteris paribus (CP) laws are a quite legitimate feature of scientific theories, some even going so far as to claim that laws of all scientific theories currently on offer are merely CP. We argue here that one of the common props of such a thesis, that there are numerous examples of CP laws in physics, is false. Moreover, besides the absence of genuine examples from physics, we suggest that otherwise unproblematic claims are rendered untestable by the mere addition of the CP operator. Thus, "CP all Fs are Gs", when read as a straightforward statement of fact, cannot be the stuff of scientific theory. Rather, we suggest that when "*ceteris paribus*" appears in scientific works it plays a pragmatic role of pointing to more respectable claims.

1. INTRODUCTION

Philosophers fall in love with arguments. As love is blind(ing), it is no surprise that an appealing argument can seduce philosophers into believing the most outlandish things. As a case in point we could cite the topic of *ceteris paribus* (CP) laws.[1] Through a constellation of arguments, very many philosophers have managed to convince themselves not only that there is such a topic but that it is an important topic meriting a never-ending stream of articles in philosophy journals. The *innamorati* are not shy of declaring their love. Thus, for example, two recent articles, Peter Lipton's (1999) "All Else Being Equal" and Michael Morreau's (1999) "Other Things Being Equal", begin with remarkably similar declarations. Lipton: "Most laws are ceteris paribus (CP) laws" (155). Morreau: "Arguably, hedged laws are the only ones we can hope to find. Laws are commonly supposed to be truths, but interesting generalizations, without some modifier such as 'ceteris paribus' are by and large false" (163).

We are bearers of bad news: put crudely, our message is that the object of their affections does not exist. To be less crude and more specific, the following seven theses are widely endorsed in the philosophical literature:

(T1)   It is legitimate for a theory of a special science (e.g., psychology, biology, economics) to posit CP laws.

Erkenntnis **57**: 281–301, 2002.
© 2002 *Kluwer Academic Publishers. Printed in the Netherlands.*

(T2)   It is scientifically legitimate for a theory of fundamental physics to posit CP laws.

(T3)   Some of our best current scientific theories (especially those in the special sciences of psychology, biology, economics etc.) posit CP laws.

(T4)   All of the laws posited by our best current scientific theories (even those of fundamental physics) are CP laws.

(T5)   There are in the world CP laws pertaining to "higher-level" phenomena such as those studied by the special sciences.

(T6)   There exist CP laws.

(T7)   All of our world's laws of nature are CP laws.

These theses span a great range of regions of philosophical inquiry: (T1) and (T2) concern scientific methodology; (T3) and (T4) concern the interpretation of current scientific theories; (T5)–(T7) concern metaphysics. But the arguments that have led our colleagues to these different theses are intimately intertwined. We maintain that although these intertwined arguments take note of important and interesting phenomena, they are deeply misleading, and that all of (T1)–(T7) are false.

Within the scope of this paper we cannot hope to set out in full the motivations for our sweeping claim. But we will attempt to convey some of the key considerations. In particular, we will analyze some of the mistakes that have led to the widely held notion that it is CP all the way down to fundamental physics, and at the same time we will set out our reasons for holding that laws are strict in fundamental physics (Section 2). Then we turn to our reasons for thinking that it is a bad idea to admit CP laws at all (Section 3). But given that the special sciences do not articulate strict laws, we are faced with the challenge of explaining the scientific status and the manifest achievements of these sciences. We will consider a way of meeting this challenge in Section 4.

While jilted lovers eventually recover, lovers of a chimera can rarely admit that their love had no object. Thus, we do not expect the *ceteris paribus* stream to dry up. But we do hope that once some of the confusions that have channelled this stream are recognized, it will take a different and more productive course.

[6]

2. IT'S NOT *ceteris paribus* ALL THE WAY DOWN

The claim (T4), which implies that it is CP all the way down for all physical laws, is a commonplace, as are the milder claims that some or most
physical laws are CP. (For example, the opening page of Morreau's (1999)
contains the assertion that, on pain of falsity, the law statements of economics, biology, and the other "non-basic sciences" must contain CP clauses.
He adds: "There are reasons to think it is so in basic sciences like physics, too" (163).[2]) (T4) has twin functions in the literature. First, it lends
legitimacy to the CP industry. The practitioners typically concentrate on
examples drawn from the special sciences, but confidence in (T4) allows
them to proceed without worry that they are focusing on some peculiar and,
perhaps, undesirable feature of the special sciences. If even fundamental
physics must resort to CP clauses when stating its laws, the thinking goes,
then surely laws qualified by such clauses are scientifically legitimate and
deserve attention from philosophers of science. Second, by rejecting the
view that among all the sciences only physics is capable of discovering
strict laws, (T4) seems to strike a blow against "physics chauvinism".

We hasten to insist that upholding "physics chauvinism" is no part
of our project. A shortcoming of much twentieth-century philosophy of
science was the assumption that physics is the paradigm science and that
other sciences are scientific only insofar as they resemble physics. Once
we give up this assumption, we should no longer automatically view any
particular apparent difference between theories of physics and theories of
other sciences as a threat to the legitimacy of the other sciences. So, given
that economics, psychology etc. evidently discover no strict laws, but at
best CP laws, it does not follow that we *must* say it is CP all the way down
in order to avoid being physics chauvinists. Moreover, we *shouldn't* say
this. The laws of representative theories from fundamental physics are not
qualified by CP clauses, as we have argued in Earman and Roberts (1999)
and Smith (2002). The claim that they are has been supported by a variety
of moves, six of which we will review and criticize in the remainder of this
section.

(i) *Appeals to examples from physics.* It is frequently alleged that actual
physical theories provide examples of CP laws. But do they really? Note
first that in order for a putative example of a "real" CP law to be interesting,
it would have to involve a CP clause that is ineliminable. The reason why
the law, as typically formulated, contains a (perhaps implicit) CP clause
whose range is not made explicit, must be that the range of this clause
cannot be made explicit. Otherwise, the CP clause is merely a function of
laziness: Though we *could* eliminate the CP clause in favor of a precise,

known conditional, we choose not to do so. There are two reasons why one might not be able to make explicit a more precise conditional: (1) we do not know how to state the conditions under which the qualified regularity holds; or (2) there is reason to suspect that even with the best of knowledge, these conditions could not be made explicit, because they will comprise an indefinitely large set. The first possibility is not really relevant here; a putative example of a CP law whose CP clause could not be eliminated just because we didn't know how to eliminate it would not show that physics actually discovers CP laws, only that it might. For all we know, future empirical research could reveal the conditions under which the regularity obtains. (Below we counter the most prominent arguments to the effect that we should not expect there to be any such conditions waiting to be revealed; see Subsections (v) and (vi).) This will be a case where what's needed is further scientific knowledge, rather than a philosophical analysis of the status of CP laws.

A physical law with a CP clause that is ineliminable for the second reason would be more interesting, and much of the literature is motivated by the belief that there are such laws (see, for example, Giere (1999) and Lange (1993, 2000)). However, it seems to us that there is no good reason to believe this, for the prominent alleged examples turn out upon scrutiny to be cases where the CP clause is eliminable. For instance, Lange claims that "To state the law of thermal expansion [which states that the change in length of an expanding metal bar is directly proportional to the change in temperature] . . . one would need to specify not only that no one is hammering the bar on one end, but also that the bar is not encased on four of its six sides in a rigid material that will not yield as the bar is heated, and so on" (Lange, 1993, p. 234). But this list is indefinite only if expressed in a language that purposely avoids terminology from physics. If one helps oneself to technical terms from physics, the condition is easily stated: The "law" of thermal expansion is rigorously true if there are no external boundary stresses on the bar throughout the process.[3] Other putative examples of indefinite conditions can likewise be easily stated within the language of physics. For instance, Kepler's "law" that planets travel in ellipses is only rigorously true if there is no force on the orbiting body other than the force of gravity from the dominant body and vice versa. Later we will argue that each of these examples is only problematically considered a law. So, they are not CP laws because (a) the CP clause is easily eliminable by a known condition, and (b) they are not laws anyway. So far, the alleged philosophical problem of CP laws has yet to make an appearance in the realm of fundamental physics.

(ii) *Confusing Hempel's provisos with ceteris paribus clauses.* Proponents of the claim that it is CP all the way down often refer to Hempel (1988) (see, for instance, Fodor (1991, p. 21), Giere (1999, pp. 90–91) and Morreau (1999, fn. 1)). However, a careful reading of Hempel's article reveals that his central concern is not the alleged need to save law statements from falsity by hedging them with CP clauses, but rather the problem of applying to a concrete physical system a theory of physics, the postulates of which are *assumed to express strict laws in no need of hedging.* Hempel notes that such an application typically requires the specification of the values of theoretical parameters, which are not ascertainable by direct observation. This simple point immediately raises a problem for the view that the empirical content of a theory is the set of its observational consequences; for if Hempel is right, this set will be null or very small. But as interesting as it is, this problem is far from the CP problem. One gets closer to the latter with Hempel's further observation that the applications of laws that physicists actually construct are often hedged. For example, a natural if somewhat crude application of Newton's theory of motion and his law of gravitation to the planets of our solar system involves assuming that this system is closed. One can, under this assumption, derive a differential equation of evolution type (or coupled set of them) that describes the motion of the planets given this assumption. The application will be valid provided that no other significant non-planetary masses are present and provided that no significant non-gravitational forces are acting on the planets. If the theory does not specify the allowed types of long-range non-gravitational forces – as Newton's original theory did not – then the second proviso has a kind of open-ended character reminiscent of CP clauses. But this does not amount to the conclusion that Newton's laws have implicit CP clauses. For in the first place, the condition for the validity of the application can be stated in precise and closed form: the magnitude of the non-gravitational forces must be small enough in comparison with the gravitational forces that the *theory* implies that the neglect of the non-gravitational forces does not affect the desired degree of accuracy of the predictions of planetary orbits. And in the second place, the conditions of the provisos are conditions for *the validity of the application,* not conditions *for the truth of the law statements of the theory*; if this were not so the theory could not be used to decide how small the magnitude of the non-gravitational forces has to be in order that they can be neglected. Once again the alleged problem of CP has failed to rear its head in physics.

(iii) *Confusing laws with differential equations of the evolution type.* We can offer here, however, a diagnosis of why it has *looked* to people as if there is a problem of CP laws in the vicinity. What makes it easy to miss the distinction between a *theory* consisting of a set of non-hedged laws and an *application of a theory* that might be hedged (though, again, in an easily stateable way) is that differential equations of evolution type – like the one we imagined deriving above – and their consequences are often thought of as *laws*.[4] If one takes them to be laws, one expects them to be *part* of the theory in question and, thus, it looks like the theory contains hedged laws. But differential equations of evolution type are not laws; rather, they represent Hempel's applications of a theory to a specific case. They are derived using (unhedged) laws along with non-nomic modelling assumptions that fit (often only approximately) the specific case one is modelling. Because they depend on such non-nomic assumptions, they are not laws. For example, because Kepler's "law" that planets travel in ellipses is derived from laws together with the assumption that there are only two bodies in the universe, it is not a law in spite of the normal nomenclature.[5] Lange's example of the "law" of heat expansion of metals is derived from a differential equation under the assumption that there are no boundary stresses, but that is a non-nomic boundary condition. The "law of free fall" is a consequence of a differential equation that involves the assumption that there is no resistance from the wind. That too is a non-nomic assumption, for it is not a law that there is no resistance from the wind. It seems to us that the role played by idealizations in physics is typically found here, in the derivation of differential equations, rather than within the laws themselves. The differential equations involve ideal-izations that need to be "hedged" in the sense described above, but this is no evidence that the laws used to derive them do.

(iv) *Early Cartwright on component forces*. In *How the Laws of Physics Lie*, Nancy Cartwright offered an argument, which still enjoys widespread influence, that special force laws like Universal Gravitation (henceforth UG) have to be merely *ceteris paribus* because they "lie" about the mo-tion of bodies. UG, for example, supposedly misrepresents the temporal behavior of an object that is also being acted upon by, say, a Coulomb force (Cartwright, 1983, 1999; Pietroski and Rey, 1995; Giere 1999). So, it must be saved from falsehood by a (usually implicit) CP claim. But, UG *cannot* misrepresent the motion of a body, because it says nothing specific about such temporal behavior.[6] Only differential equations of evolution type – which might be derivable from UG together with other consider-ations – can be integrated to describe the temporal motion of a body or system of bodies. UG cannot be so integrated. Thus, it cannot misrepresent temporal motion. In reality, what we have here is a species of the confu-sion described in the previous section: Cartwright imagines the differential equation that leaves out the Coulomb force getting the motion wrong –

which it might – and blames that on one of the laws used in deriving the differential equation, UG. But there is more packed into this differential equation than just laws. What is really wrong with the differential equation is that it was derived under the assumption that nothing carried a net charge, a false non-nomic assumption. Neither UG nor any other law forced us to assume this. Thus, the original impetus from the (alleged) falsity of UG for thinking that the special force laws are CP is ill-founded.

Cartwright is aware of the availability of this kind of objection. Her reply is that on our view, according to which special-force laws like UG do not lie about motion because they are not about motion, there is nothing left for such laws to be about. On the face of it, such laws seem to be regularities governing component forces, but according to Cartwright, there are no such forces. Cartwright's position is not a blanket anti-realism; it is a local anti-realism about component forces (which allows that, e.g., resultant forces exist). We see no viable motivation for this local anti-realism. Successful physical theories apparently quantify over component forces, and there seems to be no natural way of "paraphrasing away" reference to such forces (as there is for, e.g., references to absolute motion in Newtonian mechanics). Cartwright (1999, p. 65) has suggested that non-total forces are not "occurrent" because they are not measurable. But in the first place, in many cases they *are* measurable (e.g. a scale measures the impressed gravitational force on an object, not the *total* force on it – the latter is approximately zero, since the scale itself gives rise to a normal force that keeps the object on it from having a total acceleration downward). And in the second place it is not clear that it follows that something is not occurrent just because it is not measurable.[7]

(v) *Cartwright's argument from Aristotelian natures and experimental method.* More recently, Cartwright has defended the view that laws, including those of fundamental physics, are not regularities in behavior, but rather ascriptions of capacities to kinds of systems. She supports this view with an argument (Cartwright, 1999, Chapter 4) to the effect that two features of scientific experimental methodology are inexplicable on the view that laws describe regularities of behavior, but can be made sense of on the assumption that they are about capacities. It is not obvious that this argument is rightly characterized as an argument that it is CP all the way down to fundamental physics. Cartwright's primary goal in this argument is not to establish that all laws are CP laws, but rather to argue against a "Humean" view that restricts the ontology of science to the behaviors of physical systems and regularities in those behaviors, and in favor of a broader ontology that includes natures and capacities. Indeed, Cartwright grants that laws entail strict regularities, though these are of the form

"Systems of kind K have capacity C", rather than the form of behavioral regularities (see her essay in this issue).

The distinction between capacities and behavior obviously plays a crucial role in this view. We are not sure exactly how this distinction should be understood. It is clear that capacities are supposed to be ontologically basic posits that have an irreducible modal or causal character that is problematic for "Humeans". But many things that are naturally regarded as having such a character, such as forces, seem to fall on the behavior side of Cartwright's distinction. For, on the standard reading of Coulomb's law, it states a putative regularity among charges, positions, and forces. Cartwright insists that the correct reading of this law is not the standard one, but rather one according to which the law attributes to charged bodies the capacity to exert a force on other charged bodies. The exertion of forces, it seems, counts as behavior, whereas the capacity to exert a force does not – so the latter but not the former is the sort of thing we should expect there to be law-like regularities about.

On what may fairly be called the standard reading of fundamental physical laws, these laws do state putative regularities among behaviors. Coulomb's law, for example, states a regularity concerning the exertion of electrostatic forces among charged bodies. So, although Cartwright does not say that laws do not entail *any* strict regularities, it is fair to say that on her view, those propositions that are standardly taken to state fundamental physical laws are not true (even if our best current physical theories are true) unless qualified by a CP clause. (For Cartwright, the best way to state the CP clause is: "so long as nothing interferes with the operation of a nomological machine.") This is the claim that we deny here.

In denying this claim, we do not mean to say that we agree with Cartwright's apparent intended target. She is concerned to refute the "Humean" view of laws according to which laws just are regularities in behavior. This rather naive view, and Cartwright's view, do not exhaust the options. One can grant that there is a lot more to being a law of nature than just being a true behavioral regularity, and even grant that what laws state is helpfully understood in terms of capacities, while maintaining that laws (and capacities) must supervene on the behaviors of physical systems. For example, one could adopt something like David Lewis's (1973) best-system analysis of laws, and allow that the Lewis-laws are usefully understood as attributions of causal capacities. Cartwright, however, seems to build a lot into her notion of capacities by denying that strict regularities in behavior can be deduced from regularities about capacities alone. Her argument is intended to show that experimental methodology cannot be made sense of without supposing that the laws scientists seek to discover are claims about capacities, where these cannot be cashed out in terms of behavioral regularities (construed broadly, so that "behaviors" include such things as the exertions of forces). This is what we will try to show she does not establish.

The first feature of experimental practice that Cartwright focuses on is generalizability. A typical experiment directly tests a low-level law concerning systems of the particular kind used in the experiment. Cartwright's example is the Stanford Gravity-Probe-B experiment. What this experiment directly tests is the low-level generalization:

[A]ny fused-quartz gyroscope of just this kind – electromagnetically suspended, coated uniformly with a very, very thin layer of superfluid, read by a SQUID detector, housed in a cryogenic dewar, constructed just so . . . and spinning deep in space – will precess at the rate predicted [by the general theory of relativity]. (Cartwright, 1999, p. 88.)

But the ultimate goal of this experiment is to test a much more general claim, an "abstract" law that is part of the content of the general theory of relativity (henceforth, GTR), namely, that relativistic coupling between the spinning of a gyroscope and spacetime curvature will result in the gyroscope's precessing at a certain rate. The problem of generalizability is that of saying why what happens in *this* experiment, which concerns a system of a very specific kind, provides evidence for the more general law, which concerns systems of other kinds as well. "What is at stake is the question, 'What must be true of the experiment if a general law of any form is to be inferred from it?' " (Cartwright, 1999, p. 87).

Cartwright's answer to this question is that in a system of the kind used in the experiment, relativistic coupling is allowed "to operate according to its nature" (ibid). All other factors, whose natures involve capacities to distort the influence of relativistic coupling on precession, have been eliminated or calculated away. But when a kind of system has, by nature, a capacity to do A, and everything else with a capacity to interfere with A has been eliminated, the system will do A. Hence, in the case at hand, if relativistic coupling really has the capacity to induce precession at a certain rate, then it will do just this in the case of the Gravity-Probe-B experiment.

But it seems that Cartwright's "Humean" opponent, who believes that laws are (or supervene on) true behavioral regularities sans CP clauses, can give a similar solution. Our background knowledge includes some well-confirmed general propositions, including (the general-relativistic analogue of) Newton's second law of motion, the laws relating torque to precession, and various special-force laws. We hypothesize that there is some law relating relativistic coupling to precession. By designing the experiment in such a way that the torque-components contributed by all other known factors (such as nearby charged or massive objects, frictional properties of the material used, etc.) that are nomically related to special forces and thence to torques are very close to zero, we can make it reasonable to assume that if any precession takes place, it will all be due to relativistic coupling. Hence, we can test the prediction made by any putative law relating such coupling to precession. This procedure requires that we take for granted a great deal of background knowledge – for example, that the laws just mentioned obtain, that there really is some law relating relativistic coupling to precession, and that there aren't any other factors nomically related to precession that we have neglected to take into account. It would be fair to demand independent empirical support for each of these presumptions. However, Cartwright's solution is subject to a precisely analogous difficulty. On her analysis, the experiment presumes that the various factors relevant to the experiment really do have the capacities we take them to have; that relativistic coupling really does have some stable capacity for producing precession; that there are no capacities operative in the experimental situation at hand that have not been taken into account. So it is hard to see where Cartwright's view has an advantage over the view that laws are true regularities.[8]

Some of Cartwright's comments suggest that she thinks that unless laws were about capacities, induction wouldn't be justified. Inductive inference from a particular case to the more general case, she argues, requires that we know that the systems involved in the observed situation have capacities that remain constant and that they carry with them from one situation to another (Cartwright, 1999, p. 90; see also Cartwright, 1989, pp. 145, 157–158, 163). Otherwise, we aren't entitled to believe that there is any connection between what happens in one situation and what happens in others. But if what we need to do is justify induction, then positing capacities and natures won't help. Recall Hume's argument that assuming secret powers in nature is no help in solving the problem of induction; the problem just becomes that of explaining how we can know that the same sensible qualities are always tied to the same secret powers (Hume, 1748/1993, p. 24).

Cartwright, of course, doesn't believe what Hume seems to presume: that inductive inference in science works from the ground up. She has a bootstrapping view, according to which we always already have a store of background beliefs about natures and capacities that we can rely on in order to test new hypotheses (Cartwright, 1999, p. 98). But if the trick can be done by background beliefs about capacities, then why can't it be done by background beliefs about regularities? Why can't we justify generalizing from one experimental situation to others on the basis of background beliefs to the effect that nature is governed by laws, that these laws entail strict behavioral regularities that are true throughout spacetime, and that approximations to some of the regularities are already known?

The second prong of Cartwright's argument concerns the design of controlled experiments. In designing the Gravity-Probe-B experiment, the experimentalists had to take account of all the different kinds of factors that could influence the precession of the gyroscopes, and control each one. How did they know which factors needed to be controlled? Cartwright's answer is that they knew that everything with a capacity to influence the precession needed to be controlled. She alleges that if laws were regularities rather than ascriptions of capacities, then there would be no way for scientists to know what to control.

Why is this? Why couldn't the scientists take into account all the laws relating other factors to precession, and infer that everything related to precession by one of these laws needs to be controlled? Cartwright considers this response, and rejects it. She argues that her opponent faces a dilemma. Either the laws to be consulted are all high-level, abstract laws, or they include low-level laws about particular kinds of concrete situations. If the former, then the laws won't supply enough information to address the experimental problem at hand. This is correct, of course. If you want to know which factors need controlling in a particular experiment, it's not enough to know the most general laws of, say, electromagnetism: you also need to know the low-level laws concerning such things as the magnetization properties of the materials used in the experiment.

So the second horn of the dilemma is the one Cartwright's opponent should grasp. Where is that horn's sting? According to Cartwright, the difficulty is simply that the low-level laws are too complicated: "In low-level, highly concrete generalizations, the factors are too intertwined to teach us what will and what will not be relevant in the new design" (p. 95; see also pp. 91-92). What is puzzling about this answer is that it doesn't have anything to do with the "Humean" idea that laws are strict regularities. According to Cartwright, it is just impossible to tell which factors need to be controlled for, and which factors don't, simply because the factors are so intertwined. But that is just as true for Cartwright as it is for her opponent. In fact, the difficulty posed here is even greater for Cartwright. For according to her opponent, the contribution to the effect made by each separate factor is governed by a strict regularity, the ways in which those factors combine to produce the effect is governed by a strict regularity, and there are no other factors "peculiar to the individual case" that in principle escape the net of theory. According to Cartwright, all three of these claims are false.[9] Their falsity can only make it more difficult to determine what needs controlling and how to control it. In conclusion, the two prongs of Cartwright's argument show that there are many difficulties that experimentalists must face, but they do not show that adopting her view of laws makes these difficulties any more tractable.

(vi) *The world as a messy place.* There is one argument left for the view that it is CP all the way down: "The world is an extremely complicated place. Therefore, we just have no good reason to believe that there are any non-trivial contingent regularities that are strictly true throughout space and time". The premise of this argument is undeniably true. But it is very hard to see how to evaluate the inference from the premise to the conclusion.

Strictly speaking, this argument supports (T7), rather than (T4). As an argument for the claim that our best current theories feature only CP laws (a claim made by Cartwright (1983, 1999), Pietroski and Rey (1995), Morreau (1999)) it is impotent. Moreover, considered as an argument for (T7), it strikes us as at best an expression of despair. We will argue below that there are no CP laws, so if (T7) is true, then physical theorizing as such is a doomed enterprise. And so it might be, for all any of us know. But, in the absence of any convincing reason to think that the inference from the premise of the above argument to its conclusion is a valid one, we see no reason to surrender to despair.

3. THE TROUBLE WITH CP-LAWS

There are two important objections to the claim that CP laws play an indispensable role in science. The first is that there seems to be no acceptable account of their semantics; the second is that there seems to be no acceptable account of how they can be tested.

The first objection is the weaker one in our view, and here we only touch on it briefly. It seems that there could be no informative account of the truth-conditions of CP law-statements that did not render them vacuous. One way to see the problem is to note that we could specify the conditions under which such a statement is true if and only if we could specify the conditions under which it is false, but that is exactly what we cannot do with a CP law-statement. For such a statement will be violated exactly when the regularity contained in it is violated *and* "other things are equal", i.e. there is no "interference". But we cannot specify the conditions under which the second conjunct obtains; otherwise the CP clause is simply an eliminable abbreviation and what we have is not a genuine CP law-statement. Nonetheless, many philosophers have tried to supply truth conditions for CP law-statements (e.g., Fodor (1991), Hausman (1992)), or at least conditions for their "non-vacuity" (Pietroski and Rey, 1995).

For specific criticism of these proposals, see Earman and Roberts (1999), Schurz (2001, 2002), and Woodward (2002).

This point is not fatal to CP laws, however. Perhaps it is unreasonable to demand truth conditions for CP law-statements. This could be because the concept of a CP law is a primitive concept, which is meaningful even though it cannot be defined in more basic terms. Or it could be because an assertability semantics or conceptual-role semantics, rather than a truth-conditional semantics, is appropriate for CP law-statements. Furthermore, one might well deny that it is necessary to have an acceptable philosophical account of the semantics for a given type of statement before granting that that type of statement plays an important role in science. And it is hard to deny that there are examples of statements qualified by CP clauses that seem to be perfectly meaningful.[10]

But the second problem with CP laws, their untestability, is decisive in our view. In order for a hypothesis to be testable, it must lead us to some prediction. The prediction may be statistical in character, and in general it will depend on a set of auxiliary hypotheses. Even when these important qualifications have been added, CP law statements still fail to make any testable predictions. Consider the putative law that *CP*, all Fs are Gs. The information that x is an F, together with any auxiliary hypotheses you like, fails to entail that x is a G, or even to entail that with probability p, x is a G. For, even given this information, other things could fail to be equal, and we are not even given a way of estimating the probability that they so fail. Two qualifications have to be made. First, our claim is true only if the auxiliary hypotheses don't entail the prediction all by themselves, in which case the CP law is inessential to the prediction and doesn't get tested by a check of that prediction. Second, our claim is true only if none of the auxiliary hypotheses is the hypothesis that "other things are equal", or "there are no interferences". What if the auxiliaries *do* include the claim that other things are equal? Then either this auxiliary can be stated in a form that allows us to check whether it is true, or it can't. If it can, then the original CP law can be turned into a strict law by substituting the testable auxiliary for the CP clause. If it can't, then the prediction relies on an auxiliary hypothesis that cannot be tested itself. But it is generally, and rightly, presumed that auxiliary hypotheses must be testable in principle if they are to be used in an honest test. Hence, we can't rely on a putative CP law to make any predictions about what will be observed, or about the probability that something will be observed. If we can't do that, then it seems that we can't subject the putative CP law to any kind of empirical test.

A number of philosophers have argued that, in spite of these difficulties, CP laws can be empirically confirmed and are confirmed regularly in the special sciences (e.g., Hausman (1992) and Kincaid (1996)). Here we will not consider in detail what any single author has written about this, but will consider a couple of the most common ideas and explain why we find them unsatisfying. (More detailed criticisms of specific authors can be found in Earman and Roberts (1999).)

One common view is that we can confirm the putative law that CP, all Fs are Gs by finding evidence that in a large and interesting population, F and G are highly positively statistically correlated. Such evidence would indeed, ex hypothesi, confirm the (precise!) hypothesis that in that large and interesting population, F and G stand in a certain statistical relation. But that is not a CP law. Why confirmation of this precise claim should be taken as evidence for the truth of the amorphous claim that "CP, all Fs are Gs", from which nothing precise follows about what we should observe, has never been adequately explained. Perhaps, under certain circumstances, confirmation of this statistical claim can also lend confirmation to the stronger claim that in some broader class of populations, F and G are positively statistically correlated. That would be interesting, but again, that would not be a CP law.

It has also been suggested that we can confirm the hypothesis that CP, all Fs are Gs if we find an independent, non-ad-hoc way to explain away every apparent counter-instance, that is, every F that is not a G.[11] But this could hardly be sufficient. Many substances that are safe for human consumption are white; for every substance that is white and is not safe for human consumption, there presumably exists some explanation of its dangerousness (e.g., in terms of its chemical structure and the way it interacts with the human nervous system); these explanations are not ad hoc, but can be supported by a variety of kinds of evidence; but none of this constitutes evidence for the hypothesis that it is a law that CP, white substances are safe for human consumption. It might be complained that whiteness is not a real property, and so is unfit to appear in a law of nature. But clearly, examples like this one are easily multiplied. Substitute "compounds containing hydrogen" for "white substances", and the example works just as well.

Perhaps the testing of a putative CP law requires both explanations of apparent counter-instances, and evidence that in some large and interesting population, F and G are highly correlated. Here again, such evidence would confirm a certain set of hypotheses – (a) a hypothesis concerning the statistical relations between F and G in a certain population, and (b) one or more hypotheses explaining why many Fs are not Gs. Such hypotheses could constitute valuable empirical information. But again, they would not be CP laws. Would anything of interest be added if, in drawing our conclusions, we didn't stop with announcing the confirmation of (a) and (b), but went on to add, "And what's more: CP, Fs are Gs"? We don't see what. More importantly, we don't see what could justify or motivate this addendum. Certainly, all the evidence is accounted for by (a) and (b) alone. Further, it seems that any counterfactuals licensed by the alleged CP law could be supported by (a) and (b) as well.[12] It might be argued that the CP law provides a *good explanation* for why (a) and (b) should be true, and is thus a warranted conclusion. But it would be a supposedly explanatory hypothesis which implies no predictions over and above the evidence it supposedly explains – neither testable predictions about what we may observe, nor even predictions about unobservable features of the world.[13] The addendum would thus seem to be both empirically and theoretically otiose.

4. WHY THERE DON'T HAVE TO BE CP LAWS IN THE SPECIAL SCIENCES

It is frequently argued that many of the special sciences have managed to articulate and confirm laws of nature that can only be interpreted as CP laws. So, if one wants to deny that there are any CP laws, then one is going to have to deny the manifest achievements of the special sciences.

This argument is over-hasty. Of course it is true that many of the special sciences have made impressive achievements in describing, predicting and explaining phenomena. And it is also true that most of the apparent laws one finds in the special sciences have exceptions, and cannot be rewritten in a finite form in which they are logically contingent and exceptionless. But all that follows from this is that doing justice to the special sciences requires recognizing an important and legitimate job that can be played by CP *law-statements*. It needn't follow that there must exist propositions or facts which it is the job of CP law-statements to state. There are plenty of important things that indicative-mood sentences can do other than state propositions or facts.

To see how CP law-statements could do important work even if there are no CP laws, consider the example:

(S)   CP, smoking causes cancer.

If an oncologist claims that (S) is a law, then, we maintain, there is no proposition that she could be expressing (except for the vacuous proposition that if someone smokes, their smoking will cause them to get cancer, unless it doesn't), and even if there were, we wouldn't be able to test it. But we could come to know what she was getting at if she could say why she thought it was (S), rather than its contrary ("CP, smoking prevents cancer") that is a law. There are many things she could say about this. She could tell us that the probability of having lung cancer given that you smoke is higher than the probability of having it given that you don't. She might say that laboratory tests have shown that when certain compounds found in tobacco smoke are introduced into normal cells, they become cancerous at a higher rate than normal. And so on. The more such information the oncologist gives us, the more light dawns. Let "(I)" stand for this body of helpful information provided by the oncologist. None of the information in (I) is a CP law. It all consists of unambiguous, contingent empirical information that we know how to test using the techniques studied by standard confirmation theory.

Many philosophers of science would take (S) to be a hypothesis which is confirmed by (I). An alternative is to view the relation between (S) and (I) not as the relation between hypothesis and evidence, but rather that between an elliptical and imprecise expression of a large and unwieldy body of information, and a longer but more precise statement of that same body. On this view, the information adduced by the oncologist is indeed her reason for saying, "It is a law that, other things being equal, smoking causes cancer", but it is (part of) her *pragmatic* reason for producing a CP law-statement, rather than her *epistemic* reason for believing in the existence of a CP law.

There are other options as well. One could give a non-cognitivist account of CP law-statements, according to which the speech act of uttering such a statement is a way of expressing something which is not thereby asserted. For example, to say, "CP, all Fs are Gs" might be a way of simultaneously: (i) asserting that a great deal of precise empirical information has been gathered of the sort adduced by the oncologist in the above example; and (ii) expressing (but not asserting) that the speaker is committed to a research program that aims to explain all or most Gs in terms of Fs (together, perhaps, with other factors). The proposal is not that the CP law-statement is equivalent to the conjunction of (i) and (ii); if it were, then it would follow that if (ii) were false, then so would be the CP law-statement – but surely, anyone who asserts a CP law-statement does not mean to be asserting a proposition whose truth depends on her own commitments. Rather, the proposal is that a token utterance of a CP law-statement is a speech act in which (i) is asserted, and the information that (i) is true is pragmatically conveyed – just as a token utterance of "It is raining" pragmatically conveys the information that the speaker believes that it is raining, without asserting it. One advantage of this proposal would be that it would allow our oncologist to come to reject (S) even while continuing to maintain (I) – as she might do if she came to believe that the correlations noted in (I) were true but explicable in terms of a common cause that screens off the influence of smoking on cancer.

No doubt, there are other possible avenues for developing a pragmatist or non-cognitivist account of CP law-statements, without admitting the existence of any such proposition or fact as a CP law.[14] Pursuing some such strategy would have a number of important advantages. First of all, the strategy upholds the overwhelming appearance that the very idea of "CP laws" is either confused or vacuous, and does not (disastrously) require them to be empirically confirmable. Further, it recognizes a sense in which CP law-statements can be useful and important, and a sense in which they stand for important scientific achievements. It makes understanding the significance of any given CP law-statement a matter of knowing something about the details of current work in the science from which it comes, rather than a matter of doing philosophical analysis, or logic-chopping, on the "CP" clause. Most importantly, its availability shows that those who deny the existence of CP laws need not denigrate the achievements of the special sciences. Here we won't defend any particular version of this general strategy, which can be characterized by the slogan, "CP law-statements without CP laws". But it seems to be a hopeful strategy, whereas we have argued that the view that nature contains CP laws and science can discover them is hopeless.

5. CONCLUSION

Those enamored of CP laws typically assume that (1) the special sciences are incapable of establishing strict laws. They further assume that (2) to count as a genuine science, a discipline must be able to provide scientific predictions and scientific explanations of the phenomena in their respective domains. They also assume that (3) scientific predictions and explanations must be based on laws. And finally they assume that (4) the special sciences are sciences. They conclude that there must be CP laws and that the special sciences are capable of establishing them. Since this train of reasoning is valid and since we reject the conclusion, we must reject at least one of the four assumptions. We do not wish to question the fourth assumption. Earman and Roberts (1999) not only accepted the first assumption but gave an argument for it. Elliott Sober (private communication) has convinced us that our argument is vulnerable. Nevertheless, we continue to think that something in the neighborhood of the first assumption is true for most of the special sciences, at least insofar as they resist explicit reduction to physics. Of the remaining assumptions, our hunch is that the third is the most vulnerable and that it is worth devoting some effort to developing non-law based accounts of explanation and prediction.[15] The second assumption also deserves some critical examination, but an outright rejection of this assumption would bring into question the standard ways of trying to demarcate the genuine from the pseudo-sciences.

In sum, the way ahead is not clear. But what is clear is that our rejection of the entrenched views on CP laws has important ramifications for the philosophy of the special sciences. In particular, it points to a kind of pluralism. There is an important difference between fundamental physics and the special sciences with regard to the laws they discover and the forms of explanation they can produce. This, however, is no threat to the legitimacy of the special sciences.

NOTES

[1] In what follows we will use both "CP law statement" and "hedged law statement" to mean a statement of the form "CP (i.e. all other things being equal) Φ" where Φ is a strict law statement, i.e. a non-hedged proposition asserting a lawlike generalization of either universal or statistical form. A CP law is a law supposedly expressed by a CP law statement. The distinction between CP law statements and CP laws will become important in Section 4.

[2] The claim is typically allowed to stand unchallenged. To our knowledge, the only sustained attacks are found in Earman and Roberts (1999) and Smith (2002).

[3] Of course, it will be close to true even when there are some external stresses – as there usually are – but they are small. Deviations from the "law" for small stresses should be easily calculable and are not essentially problematic.

[4] A differential equation of evolution type is a differential equation with time as the independent variable, describing the evolution in the physical magnitudes of a given system over a given stretch of time.

[5] It is also not a law for the reason that the ellipses only result from the differential equation for some initial conditions, and initial conditions are generally taken to be non-nomic.

[6] This is obvious in that UG does not even involve time as a variable at all.

[7] Lange (this issue) suggests a plausible reason for denying the reality of the component electrostatic forces apparently governed by Coulomb's law. As Lange notes, the standard argument for the reality of the electric field does not carry over to the reality of the components of this field contributed by individual charged bodies. However, carrying over the standard argument for the reality of the electric field is not the only way to argue for the reality of component electrostatic forces. For example, in some cases, such forces can be measured; a torsion balance can be used to measure an electrostatic force by measuring the torsion force needed to counterbalance it.

[8] Of course, the "Humean" line just sketched assumes that component forces, such as the electrostatic force exerted on one body by another, exist, and that special-force laws are true regularities concerning such forces. Cartwright famously rejects these assumptions.

For example, she dismisses the very idea of force due to charge as "no concept for an empiricist". We find this a peculiar argument in this context. In the same chapter, Cartwright acknowledges that she herself has "made the empiricist turn" (Cartwright, 1999, p. 81), yet she is happy to allow the existence of *capacities* possessed by systems *due to their natures*, which persist *even in conditions where they do not manifest themselves in observable behavior*. "We no longer expect that the natures that are fundamental for physics will exhibit themselves directly in the regular or typical behavior of observable phenomena" (ibid). If natures, together with the capacities they ground, persisting in objects even when they are not exhibited in observable behavior, are acceptable to an empiricist, then we wonder what is so bad about forces due to charges, which do not typically manifest themselves as the whole resultant force responsible for observable motion, but which can be measured in certain controlled situations.

9 Concerning the third, see Cartwright (1989, p. 207).

10 For example, Lange's example concerning officers in the British navy, discussed in his paper in this volume.

11 This is the proposal of Pietroski and Rey (1995). Lange (this volume) suggests that CP laws can be testable because we can find genuine counterexamples to them, by finding counter-instances that are clearly not covered by the CP clause and can only be excused in an ad hoc way. It seems to us that our response to Pietroski and Rey applies to Lange's proposal as well, though it would take more work to show this in detail.

12 Schurz (2001) argues that in a deterministic world such explanations are always forthcoming.

13 Consider the counterfactual: "If the water in this cup had been drawn from the Atlantic Ocean, then it would be salty". One can support this counterfactual by pointing out that (a) almost all samples of water drawn from the Atlantic are salty, and (b) almost all such samples that *aren't* salty aren't salty because they have been subjected to a purifying process (which we know that the water in this cup has not been). (If this water *has* been subjected to such a process, then the counterfactual is presumably false.) There is no need to back up the counterfactual by alleging that it is a law of nature that, CP, water drawn from the Atlantic is salty. It is plausible that counterfactuals supported by alleged CP laws could also be supported in ways similar to the one just illustrated.

14 Lange (2000) articulates and defends a sophisticated theory according to which laws are not regularities at all, but rather encode rules of inference that belong to "our best inductive strategies". One of the advantages Lange claims for his account is that it handles CP laws very naturally, since a rule of inference might fail to be truth-preserving in all cases yet still from part of a very good strategy for studying the world. Two of us (Earman and Roberts, 1999, pp. 449–451) have criticized Lange on this issue. However, we now think that Lange's account might be exactly right for the case of *CP* laws. It is a "non-cognitivist" account of laws in the sense that it takes laws to express normative features of our scientific practice rather than to describe the (natural, non-normative) features of the world. We disagree with Lange, however, on the topic of the laws of fundamental physics.

15 Here we applaud the efforts of Woodward (2000, 2002).

REFERENCES

Cartwright, N.: 1983, *How the Laws of Nature Lie*, Oxford University Press, Oxford.
Cartwright, N.: 1989, *Nature's Capacities and Their Measurement*, Oxford University Press, Oxford.

Cartwright, N.: 1999, *The Dappled World: A Study of the Boundaries of Science*, Cambridge University Press, Cambridge.

Cartwright, N.: 2002, 'In Favor of Laws That Are Not *Ceteris Paribus* After All', (this issue).

Earman, J. and J. Roberts: 1999, '*Ceteris Paribus*, There Are No Provisos', *Synthese* **118**, 439–478.

Fodor, J.: 1991, 'You Can Fool Some of the People All the Time, Everything Else Being Equal: Hedged Laws and Psychological Explanations', *Mind* **100**, 19–34.

Giere, R.: 1999, *Science without Laws*, University of Chicago Press, Chicago.

Hausman, D.: 1992, *The Inexact and Separate Science of Economics*, Cambridge University Press, Cambridge.

Hempel, C. G.: 1988, 'Provisos: A Philosophical Problem Concerning the Inferential Function of Scientific Laws', in A. Grünbaum and W. Salmon (eds), *The Limits of Deductivism*, University of California Press, Berkeley, CA, pp. 19–36.

Hume, D.: 1748/1993, *An Enquiry Concerning Human Understanding*, Oxford University Press, Oxford, Cambridge.

Kincaid, H.: 1996. *Philosophical Foundations of the Social Sciences*, Cambridge University Press, Cambridge.

Lange, M.: 1993, 'Natural Laws and the Problem of Provisos', *Erkenntnis* **38**, 233–248.

Lange, M.: 2000, *Natural Laws in Scientific Practice*, Oxford University Press, Oxford.

Lange, M.: 2002, 'Who's Afraid of *Ceteris Paribus* Laws? Or: How I Stopped Worrying and Learned to Love Them', (this issue).

Lewis, D.: 1973, *Counterfactuals*, Harvard University Press, Cambridge, MA.

Lipton, P.: 1999, 'All Else Being Equal', *Philosophy* **74**, 155–168.

Morreau, M.: 1999, 'Other Things Being Equal', *Philosophical Studies* **96**, 163–182.

Pietroski, P. and G. Rey: 1995, 'When Other Things Aren't Equal: Saving *Ceteris Paribus* Laws from Vacuity', *British Journal for the Philosophy of Science* **46**, 81–110.

Schurz, G.: 2001, 'Pietroski and Rey on *Ceteris Paribus* Laws', *British Journal for the Philosophy of Science* **52**, 359–370.

Schurz, G.: 2002, '*Ceteris Paribus* Laws', (this issue).

Smith, S.: 2002, 'Violated Laws, *Ceteris Paribus* Clauses, and Capacities', *Synthese* **130**(2), 235–264.

Woodward, J.: 2000, 'Explanation and Invariance in the Special Sciences', *British Journal for the Philosophy of Science* **51**, 197–254.

Woodward, J.: 2002, 'There Is No Such Thing as a *Ceteris Paribus* Law', (this issue).

John Earman
Department of History and Philosophy of Science
University of Pittsburgh
Pittsburgh, PA 15217
U.S.A.
E-mail: jearman@pitt.edu

John T. Roberts
Department of Philosophy
University of North Carolina at Chapel Hill
Chapel Hill, NC 27599
U.S.A.

Sheldon Smith
Metropolitan State College of Denver
Denver, CO 80217-3362
U.S.A.

JIM WOODWARD

THERE IS NO SUCH THING AS A *CETERIS PARIBUS* LAW

ABSTRACT. In this paper I criticize the commonly accepted idea that the generalizations
of the special sciences should be construed as *ceteris paribus* laws. This idea rests on mistaken assumptions about the role of laws in explanation and their relation to causal claims.
Moreover, the major proposals in the literature for the analysis of *ceteris paribus* laws are,
on their own terms, complete failures. I sketch a more adequate alternative account of the
content of causal generalizations in the special sciences which I argue should replace the
ceteris paribus conception.

1.

The so-called problem of *ceteris paribus* (hereafter cp) laws is one of those
cases – unfortunately common in philosophy – in which interesting and important issues have become enmeshed in a framework that interferes with
their constructive exploration. As I understand it, the dialectic surrounding
the problem of cp laws goes something like this: Many philosophers hold
the following set of beliefs. (1) A genuine science must contain "laws". (2)
Whatever else a law is, it must at least describe an exceptionless regularity.
In particular all laws have the "All As are Bs" form of (U) universally
quantified conditionals in which the condition in the antecedent of the
law is "nomically sufficient" for the condition in its consequent. (3) Laws
are required for successful explanation and to ground or support causal
claims. Even if the DN model of explanation didn't quite get the details
right, explanation is at bottom a matter of providing nomically sufficient
conditions for an explanandum and this requires generalizations that are
laws. (4) Putting aside generalizations that are explicitly probabilistic in
form, if a generalization is to be testable at all (if it is to have empirical
content rather than being vacuous), it must take the form (U). If it does
not, we cannot use the generalization to make determinate predictions.

These views raise an immediate problem when we confront them with
the generalizations of the special sciences, few of which seem to be exceptionless or of form (U). One possible response is that this shows that
the special sciences are not really sciences and that they largely employ
generalizations that cannot figure in explanations and are untestable. Most

Erkenntnis **57**: 303–328, 2002.
© 2002 *Kluwer Academic Publishers. Printed in the Netherlands.*

philosophers have been unwilling to accept this response. Instead, the most common strategy has been to continue to accept beliefs (1)–(4), but to search for a way of construing the generalizations of the special sciences as "laws", despite appearances to the contrary. It is this strategy which motivates the *ceteris paribus* laws literature. The idea is that the generalizations of the special sciences, despite failing to state (at least explicitly) nomically sufficient conditions for outcomes, nonetheless (at least sometimes or if appropriate conditions are met) can be regarded as a special kind of law – a *ceteris paribus* law. Because they are laws, cp laws can figure in explanations and are testable. Thus the scientific status of the special sciences is vindicated. The project for enthusiasts for cp laws thus becomes one of specifying the conditions under which a cp law is, in the language of Earman and Roberts (1999), "scientifically legitimate" (true, non-vacuous, testable, supported by evidence, capable of figuring in explanations etc.)

On my view, this entire enterprise is misguided. First, and most fundamentally, each of the motivating assumptions (1)–(4) is wrong-headed. It is false that to qualify as genuine science a discipline must contain laws. Among other things, the notion of a law of nature is not sufficiently clear and the borderline between law and non-law too hazy for it to play this sort of demarcative role.[1] It is also false that successful explanation requires laws and false that the provision of a nomically sufficient condition for an explanandum is either necessary or sufficient for explaining it. Finally, the argument about testing sketched under (4) above is misguided for many reasons, the most immediately relevant of which is that relies on an overly restricted view of what can be predicted from a generalization. Because assumptions (1)–(4) are misguided, there is, as far as I can see, no motivation for the whole cp laws enterprise, understood as the project of construing the generalizations of the special science as laws of a special sort and then searching for general conditions for them to be legitimate.

Nor is this the only problem with the cp laws literature. A second fundamental difficulty – one that ought to carry weight even with those who do not share my conviction that the motivational assumptions that guide the enterprise are mistaken – is that the major proposals in the literature for the analysis of cp laws are, on their own terms, complete failures. Moreover, the pattern of failures makes it hard to believe that the analyses are fundamentally on the right track, correctly capturing core cases, but breaking down when applied to devious, unusual counterexamples. Instead, the analyses fail quite systematically – they don't return the right answers even in core cases. I believe that this systematic pattern of failure derives from the falsity of the motivational assumptions that guide the cp project.

There are other reasons as well to be skeptical of the notion of a cp law. Although many philosophers seem to be under the impression that gen-ʹ eralizations that explicitly incorporate *"ceteris paribus"* clauses or other qualifying expressions of similar indeterminacy but different meaning ("if no interfering or disturbing factors are present" etc.) are common in the special sciences, it seems to me that this is simply not the case. To the best of my knowledge the only discipline in which the *"ceteris paribus"* locution itself is explicitly used with any frequency is economics, where it has a very specific meaning that does not readily generalize to other contexts. The idea that the generalizations of the special sciences should be regarded as incorporating *ceteris paribus* or other qualifying clauses is a philosopher's gloss on how these generalizations should be understood, and not an idea that draws any support from the way in which those generalizations are actually formulated by the researchers who use them. One consequence is that it is often unclear what in the special sciences corresponds to the notion that is supposedly reconstructed in the cp laws literature. This makes it hard to judge the adequacy of those reconstructions.

A closely related point concerns the great diversity and heterogeneity of the generalizations that philosophers propose to analyze in terms of the category of *ceteris paribus* laws. Some of these, like the generalization (E) considered below, which tells us about the effect of certain chemotherapy drugs on tumor remission, explicitly use words like "cause" and tell us about an effect produced by some causal factor, but do not have the form of deterministic generalizations, it being understood instead that the cause will produce the effect only when certain other circumstances, not specified in the generalization, are present. Other generalizations that are taken by philosophers to have an implicit *ceteris paribus* clause attached to them are most naturally understood, not as claims about the overall or net effect that will occur when other conditions are present, but rather as generalizations about some component or feature of the effect that is attributable to the operation of specified set of causal factors, when these are taken by themselves or are conceived as operating in isolation. For example, the gravitational inverse square law (which on my view should be understood as describing the gravitational component of the total force experienced by a mass) is sometimes claimed to be implicitly qualified by a *ceteris paribus* clause (referring to the absence of non-gravitational forces) since (it is argued) it is incorrect when non-gravitational forces are present (Cartwright, 1983; Hausman, 1992).[2] Still others, like the ideal gas law $PV=nRT$, are (I would argue) causal in character, but unlike (E) have the form of deterministic generalizations, even though it is common knowledge that they break down or have exceptions in various circumstances and may hold only approximately even in the cases to which they are usually applied. Still other generalizations (such as the generalization that most mammals have hair) that are sometimes treated in the literature as cp laws tell us that some regularity or uniformity holds usually or in most cases, but (so I argue – see below) carry no suggestion that the regularity is causal. Although there seems to be a presumption in the cp literature that some single analysis will apply to all or most of these various kinds of examples, I see no reason to assume this. My contrary view is that one of the many reason(s) for the systematic failure of the various proposed analyses of cp laws is that they attempt to capture generalizations with very different content by means of a single analysis.

While I reject the idea that the generalizations found in the special sciences are *ceteris paribus* laws, I fully agree that many of those generalizations are "scientifically legitimate", that they are testable and in fact strongly supported by evidence, and that they describe causal relationships and figure in explanations. My claim is that construing those generalizations as *ceteris paribus* laws is the wrong way to defend their usefulness and legitimacy and casts no real light on how they can be supported by evidence or used to explain. My disagreement with the advocates of cp laws is thus *not* merely (or even primarily) a "humpty-dumpty" type disagreement (Glymour, this volume) about how to use the word "law". Instead, as I see it, the real problem with the *ceteris paribus* solution is that directs our attention along an unproductive avenue – it encourages us to believe that we can vindicate the special sciences simply by coming up with the right account of the semantics of *ceteris paribus* locution, while retaining in an unaltered form the accounts of law, testing and explanation associated with (1)–(4). I claim that what is required instead is the hard work of rethinking (1)–(4) and replacing them with more adequate accounts of testing and explanation in the special sciences.

2.

Before turning to details, it will be useful to have a concrete example in front of us. Over the past few decades researchers have discovered a number of chemotherapy drugs that are sometimes effective against different forms of cancer. Discovery and testing of these drugs largely proceeds on an empirical, case-by-case basis. While there is often a general theoretical understanding of the mode of action of various drugs – e.g., that a particular drug works by interfering with DNA replication or cell signaling and does this in a way that affects tumors more strongly than normal tissue, there is no framework available at present that allows one to predict from more fundamental biological theory which drugs will be efficacious and which will not. Moreover, even drugs that are efficacious against certain tumors in some individuals are not efficacious against similarly classified tumors in other individuals. It is widely believed that one important reason for this is that tumors that appear similar under a microscope, and accordingly are classified by pathologists as belonging to the same type, differ from one another at a molecular level and in their genetics and that these differences are important to how the tumors respond to chemotherapy agents. In most cases, however, researchers cannot at present predict from molecular or genetic analysis of a tumor whether it is likely to respond favorably to a given drug, although it is widely anticipated that much more will be learned about such questions in the coming decade. In addition, of course, response to chemotherapy may be influenced by many other individual factors such as patient metabolism, age, general health and so on. Thus, whether an individual patient's tumor will respond favorably to a given drug can often only be discovered by administering the drug to the patient and seeing whether it works – whether the tumor shrinks or disappears.

There are further complications surrounding the notion of administration of a drug. Each drug is usually administered in concert with a suite of other chemotherapy drugs, according to a specified protocol – it is really the effectiveness of the whole protocol that is tested in experimental trials of the drug. The protocol fixes certain details of the administration of the drugs but makes others contingent on the results of various measurements made on the patient. For example, a protocol might specify that three rounds of a certain combination of drugs are to be administered at intervals between fourteen and twenty eight days, depending on when (and whether) various measurements of the patient's white cell count, liver functioning etc. are at acceptable levels. As a consequence, "administration A of complex of drugs D according to protocol P" almost certainly picks out a causally heterogeneous class of therapeutic interventions, in the sense that different instances of A (e.g., same drugs and same protocol, but one involving three rounds at fourteen day intervals, and another involving three rounds at twenty eight day intervals) may have a different causal impact on otherwise identical patients.

Let us consider the claim that

(E)   Administration A (where this consists administration of drugs D according to protocol P) to human beings with tumors of type T causes recovery R (where recovery is understood to mean that the tumors of type T disappear and remain absent for some specified period – e.g., five years).

[31]

For some values of A, T, and R, (E) is true, widely believed to be true, and strongly supported by evidence. For example, treatment with the drugs cisplatin and vincristine, in concert with other drugs and radiation, is effective against some but not all supratentorial primitive neuroectodermal tumors (see, e.g. Reddy, Janss, Philips, Weiss, and Hacker, 2000). Let us stipulate that we are dealing with such a case. How should (E) be understood? Consider the following line of thought. The general form of a law of nature is

(U)   All As are Bs.

If we try to put (E) into form (U) we get something like

(E*)   All human beings with tumors of type T who undergo A will recover.

However, while (E) is true, (E*) is false, for reasons described above. Hence (E*) doesn't capture the content of (E). What to do? The first step in the strategy proposed by fans of *ceteris paribus* laws is to construe (E) instead as a *"ceteris paribus"* version of (E*) – that is, as the claim that

(CPE)   *Ceteris paribus*, all human beings with tumors of type T who undergo A will recover.

We are now, however, faced with the question of what (CPE) means and how it can be tested. In particular, what conditions must the cp clause in (CPE) meet if (CPE) is to be "scientifically legitimate"? Pietroski and Rey (1995), in their recent attempt to specify the conditions under which (as they put it) cp laws are "non-vacuous", capture the central difficulty very concisely:

There is a legitimate worry that appeals to cp-clauses render the nomic statements they modify somehow vacuous or unacceptably circular. In particular, if 'cp, F → G' means merely that 'F → G' is true in those circumstances in which there are no instances that are F and not G, then 'c.p.laws' look to be strictly tautologous – true, but presumably not explanatory laws in an empirical science. (p. 87)

With a few exceptions the advocates of *ceteris paribus* laws have pursued the same generic strategy for resolving this difficulty. The starting point is the assumption that so-called strict or universal laws (i.e., laws of form (U) that explicitly state nomically sufficient conditions for the outcomes described in their consequents) are unproblematically legitimate. It is further assumed that *ceteris paribus* laws will be legitimate to the extent that they can be regarded as stand-ins for underlying strict laws or to the extent that the *ceteris paribus* clauses in them can be "discharged" in a way that connects them in some appropriate way to such underlying laws.

As an initial illustration of this strategy let us consider the claim that it is at least a necessary condition for a cp law of the form "*Ceteris paribus*, all Fs are Gs" to be legitimate, that there be (i) some further condition K such that "All Fs which are K are G" is a strict law. As it stands condition (i) will be trivially satisfied if the domain of interest is deterministic, since under that assumption there will always be some condition (which may have nothing to with F) which will be nomically sufficient for G. (Moreover, there would be no motivation for (i) in the first place if the domain were not deterministic, since under indeterminism there is no reason to suppose that there is such a condition K.)[3] To modify slightly an example from Earman and Roberts (1999) (i) is satisfied for "*Ceteris Paribus*, all spherical bodies are conductors" since there is some set of conditions that are nomically sufficient for a body, spherical or otherwise, to be a conductor.

We can avoid this particular problem if we follow Fodor (1991) by adding two additional conditions to (i):

(ii) F is not by itself nomically sufficient for G.
(iii) K is not by itself nomically sufficient for G.

Modifying slightly the terminology in Fodor (1991), let us call a condition K that meets (i)–(iii) a *completer* for F with respect to G.

We are now in a position to consider (cleaned up versions of) some of the proposals in the literature regarding the conditions that must be met for a generalization to qualify as a legitimate cp law. One natural suggestion is:

> **(F2)**   "All Fs are Gs" is a *ceteris paribus* law if and only if there exists a completer K for F with respect to G.

(**F2**) is similar to but not identical with the proposal that Fodor calls "Story 2" about *ceteris paribus* laws in his (1991).[4] Two other proposals along broadly similar lines are Hausman (1992) and Pietroski and Rey (1995). Hausman requires merely that the "*ceteris paribus* clause picks out some predicate that when added to the antecedent of the unqualified generaliz-ation makes it an exact [i.e., strict] law" (1992, p. 137). As it stands, his proposal is thus subject to the trivialization described above in connection with (i). However, a more charitable reading is that Hausman takes the satisfaction of (i)–(iii) to be necessary and sufficient for "All Fs are Gs" to be a *ceteris paribus* law. On this interpretation, Hausman is also committed to (**F2**).

Pietroski and Rey (1995) require that for "All Fs are Gs" to be a *ceteris paribus* law, then either instances of F are either followed by G or in cases in which instances of F are not G, there is some factor K′ which explains why this is the case and which also explains other facts as well. Pietroski and Rey also do not explicitly impose (ii)–(iii) but I take a charitable reading of their proposal to be (something like):

(PR)   "All Fs are Gs" is a *ceteris paribus* law if and only if there exists a completer K for F with respect to G, and when Fs are not G, some condition K′ obtains such that all F and K′ are not G and such that K′ does some additional independent explanatory work, in the sense that in addition to figuring in the explanation of why Fs are not Gs, K′ also explains other facts as well.

While both **(F2)** and **(PR)** avoid the counterexamples in Earman and Roberts, 1999, they are nonetheless far too permissive.[5] On both proposals, the generalization "All charged objects accelerate at 10 m/s²" qualifies a *ceteris paribus* law. For every charged object, there is an additional condition K (having to do with the application of an electromagnetic field of appropriate strength to the object) that in conjunction with the object's being charged is nomically sufficient for its accelerating at 10 m/s². Moreover, neither being a charged object by itself nor being an object to which an electromagnetic field is applied is nomically sufficient for undergoing this acceleration. Thus both (ii) and (iii) are satisfied and K is a completer. Moreover, for those charged objects that do not accelerate at 10 m/s², there is always an explanation that appeals to some other factor K′ for why this is so – K′ will presumably have to do with the fact that the object in question has been subjected to an electromagnetic field (or some other force) of the wrong magnitude to produce this acceleration. In addition, since classical electromagnetism is a powerful, non ad hoc theory, K′ will figure in the explanation of many other facts – the trajectories of other particles in the field etc. So **(PR)** is satisfied as well. Even more alarmingly, parallel reasoning can be used to show that "All charged particles accelerate at *n* m/s²" is a *ceteris paribus* law for all other values of *n*. Similar reasoning shows that both **(F2)** and **(PR)** also yield the result that "All human beings {speak English with a southern accent, speak French, speak in baby talk , have vocabularies of between one and two hundred words" etc.} are *ceteris paribus* laws.[6]

One possible response to these difficulties is to add another clause to the analysis, requiring that in addition to satisfaction of the conditions in **(F2)** or **(PR)**, the following condition must hold:

(M)   If all Fs are Gs is a cp law, Fs must "usually" or "normally" be followed by Gs, or "All Fs are Gs" must hold in "most of the intended applications" of this generalization.

Related condition are advocated by a number or writers, including Earman and Roberts (1999) who offer the "most of the intended applications" condition as part of an explication of what they call the "strong sense" of "*Ceteris paribus*, all Fs are Gs". Putting aside the vagueness of "usually", "most intended applications" etc., the addition of (**M**) will exclude the counterexamples described in the previous paragraph. Nonetheless, it is arguable that (M1) "All human beings living in East Asia speak Chinese", (M2) "All people living in Western Europe speak Indo-European languages", as well as (M3) "Drivers in England drive on the left" will satisfy (**M**) as well as (**PR**) and (**F2**) and hence will still qualify as *ceteris paribus* laws.

The addition of (**M**) to the analysis also raises a deeper problem, which has already been alluded to – namely that it is unclear what in actual scientific practice is supposed to correspond to the notion of a cp law and hence also unclear how to judge the adequacy of the proposed explication. One can of course introduce "*ceteris paribus* law" as a technical term, stipulating that this simply means a generalization that holds in "most" etc. cases and meets certain additional conditions. However, if (as one would have thought) the underlying motivation of the *ceteris paribus* laws literature is to formulate conditions that are necessary and sufficient for generalizations in the special sciences to be testable, true, figure in explanations etc., then examples like (E) above, describing the effect of chemotherapy agents on tumors, seem to show that conditions like (**M**) are misguided. As I will argue below (and is perhaps already obvious) a generalization like (E) will be testable, non-vacuous, and can figure in explanations even if only a small portion of those who are treated recover, as long as in properly designed experiments the incidence of recovery is higher among those who are treated than those who are not. It may indeed be that in such a case, many philosophers would be reluctant to describe (E) as a *ceteris paribus* "law" but why doesn't this show that "*ceteris paribus* law" is a bad category to use if we want to understand generalizations like (E)? A similar point seems to hold for (M1)–(M3). These generalizations are mere descriptive summaries of uniformities or regularities in peoples' behavior – uniformities that investigators should seek to explain, rather than uniformities to which they might appeal in explaining behavior. They are not the sorts of generalizations that serious social scientists would regard as "laws" or as components of an explanatory theory or even as "causal". Again, one can simply stipulate that "cp law" means "generalization describing a uniformity that holds for the most part etc.", but the question of whether this is a useful category for understanding the special sciences and whether such generalizations can play anything like the role traditionally assigned to laws remains.

As a further indication that there is something fundamentally wrong with all of the proposals canvassed above, consider

(V)   All Fs are Gs, except when they are not.

(V) is, one would have thought, a paradigm case of a vacuous generalization. It is exactly the possibility that "*Ceteris paribus*, all Fs are Gs" is equivalent to (V) that the various accounts of cp laws considered above are designed to exclude. It is thus a source of concern that under some additional assumptions that seem to make (V) no less vacuous, it turns out to be an acceptable *ceteris paribus* law on all of the above accounts. To see this, let us suppose that some, but not all Fs are Gs and that the system under discussion is deterministic. Then there will be some condition K that in conjunction with F is nomically sufficient for G. Moreover, it is certainly consistent with (V) and these assumptions to assume also, in accord with (ii) and (iii) above, that neither K nor F by itself is sufficient for G. Under these conditions, the requirements in (**F2**) are satisfied. It is also consistent with these assumptions to assume, in accord with (**PR**), that there is some additional condition K′ that in conjunction with F can be used to explain why Fs which are not Gs occur and which also figures in the explanation of other facts – i.e., that (V) qualifies as a *ceteris paribus* law according to (**PR**). But intuitively the satisfaction of these assumptions has no tendency at all to make (V) any less vacuous.

These observations point to what seems to me to be a fundamental confusion in proposals like (**F2**) and (**PR**). Quite apart from the difficulties described in previous paragraphs, whether the conditions in (**F2**) and (**PR**) are satisfied by some generalization G cannot by themselves be what determines whether G is non-vacuous, testable or otherwise legitimate, independently of what we know about those conditions or whether we have the ability to independently identify whether they are satisfied or what we intend or mean when we use G. It may well be that in the case of (V) there is a condition K that is a completer for F with respect to G and it may also be the case that there is some condition K′ that explains in a non ad hoc way when Fs fail to be G. However, if, as is the case, (V) itself conveys no information about what those conditions are or how to independently identify when they obtain and we have no information about these matters from any other source, it is hard to see how the mere fact that these conditions hold makes (V) "legitimate".[7]

It is true enough that if we were to *replace* (V) with

(V*)   All Fs which are K are G and all Fs which are not K′ are not G.

and if we knew how to specify the conditions K and K' in an independent substantive way, then (V*) would be non-vacuous, testable etc. However, it doesn't follow from this observation that (V) itself is non-vacuous.

A similar point holds in connection with (**M**). Suppose that it happens to be true that most Fs are Gs although no one knows this or has no way of specifying the conditions under which Fs are or are not Gs. Then (V) will satisfy (**M**) but again it does not seem to follow that (V) is legitimate.[8]

3.

These problems with standard treatments of cp laws suggest that there is at least some reason exploring an alternative account of generalizations like (E). (Administration A of chemotherapy drugs causes recovery R). Let us begin by asking how one would go about testing such a generalization. (Here I will be concerned with in-principle considerations and will neglect difficult questions surrounding the ethics of various sorts of clinical trials.) One possibility is to carry out a randomized experiment: subjects with tumors of type T are randomly assigned to either a treatment group all of whom receive A or to a control group that receives no drug with chemo-therapeutic properties. The intent of the randomization is to insure that any other factors besides A that affect whether the tumor disappears (genetic differences among tumors etc.) are roughly evenly distributed between the two groups, so that any difference in the incidence of recovery between the two groups will be attributable to A. Note that randomization allows us to do this (at least in principle) even if we don't know what these additional factors are and/or are unable to ascertain whether they are present or absent in particular individuals, independently of their response to treatment. It thus allows us to test (E) even if we do not know how to specify a completer for (E) or even whether there is such a completing condition and even if we don't know how to independently explain failures to recover despite having been given A or even whether there is such an independent ex-planation. In other words, information about whether conditions like (**F2**), (**PR**) or (**M**) are satisfied plays no obvious role at all in the randomized experiment just described, which is just what we would expect, since as we have seen, satisfaction of those conditions appears to be irrelevant to whether a generalization is of a sort that we would intuitively regard as a cp law or descriptive of a causal relationship.

Suppose that the incidence of recovery R in the treatment group is considerably higher than in the control group. Standard statistical tests allow one to assess the null hypothesis that both groups can be regarded as random draws from a larger population in which there is no difference in

R. If one can reject this hypothesis and one finds a systematically higher incidence of recovery in the treatment group in other similar experiments, one has very good reason to believe that (E) is true.[9]

This experiment draws our attention to features of (E) that are missed or at least insufficiently emphasized in the reconstructions (like (**F2**), (**PR**) and (**M**) above) that are considered in the cp literature. These reconstructions begin with the assumption that (E) is problematic because it fails to state conditions which in conjunction with the *presence* of A are nomically sufficient for recovery and then attempt to add additional requirements having to do roughly with the conditions under which the presence of As always leads to R (and perhaps also, as with (**PR**), conditions under which A leads to not R) that together insure that (E) is non-vacuous etc. However, many of the generalizations of the special sciences (including (E)) do *not* just tell us what happens in the presence of some property F; they also tell us what will happen in the *absence* of F or, if F is a variable, under *changes* in the value of F. For example, (E) does not just make a claim about what happens with respect to recovery when administration A of the drug occurs. It also tells us (or implies) that a change in whether the drugs are administered to a group of subjects with tumors changes whether or not they will recover (or at least changes the probability or incidence of recovery). In particular, (E) predicts a contrast or difference in the incidence of recovery among those treated and those untreated with A, at least under the right conditions. It is this feature of (E) that we appeal to when we test it via the experiment described above. This feature is also closely bound up with our judgment that A is causally relevant to recovery.

What is the relationship between the claim that A "makes a difference to" or is "causally relevant to" R (in the sense described in the preceding paragraph, according to which whether or not A is present leads to a change in whether or not R is present or to the probability of R being present) and the various conditions for the legitimacy of cp laws canvassed above? To begin with, it is a familiar point that the fact that A in conjunction with appropriate other factors is nomically sufficient for R is not sufficient for A to be causally relevant to R. Taking birth control pills in conjunction with being a male is nomically sufficient for failing to become pregnant but taking birth control pills does not make a difference to (is not causally relevant to) whether a male becomes pregnant in the sense that there is a difference in the expected incidence of pregnancy among those males who do as opposed to those males who do not take birth control pills. Indeed, the counterexample due to Earman and Roberts described above exploits this very fact – from the fact that there are conditions K* which, in conjunction with being spherical, are sufficient for a substance's being an electrical conductor, it does not follow that being spherical makes a difference for or is relevant to being an electrical conductor, and this is at least part of the reason why we are disinclined to think of "*Ceteris paribus*, all spherical objects as conductors" is a cp law.

We noted above that we may exclude the counterexample just described by imposing condition (**F2**) – since K* by itself is sufficient for electrical conductivity, it is not a completer of being spherical with respect to electrical conductivity. This may seem to suggest that we can capture the notion of Fs making a difference for G, by appealing to the notion of a completer – F makes a difference for G if and only if there is a completer for F with respect to G.[10] However, the discussion in Section 2 shows that this strategy is unlikely to be successful in capturing any notion of "difference making" that is appropriate for an understanding of cp laws. First, there is the technical difficulty described in note 5 – even if F is causally relevant to G, it is not obvious why we should suppose that we can always formulate a K that is sufficient for G only in conjunction with F and not by itself. If, so to speak, we choose the wrong F as our starting point, this may not be possible. Second, even if it is true that a completing condition K for F with respect to G exists whenever F "makes a difference" to G, the converse of this claim does not seem to be true. One way in which the converse can fail is if the information about difference making or relevance (or at least interesting or important information about this) is not conveyed by F itself but is rather buried in the unknown completer. To modify slightly an example used above, the generalization (3.1) "all masses of 5 kg accelerate at ten meters/s^2" has a completing condition, since for each such mass there is a force that produces this acceleration and it is the conjunction of the force and the mass that produces the acceleration, not the force or the mass by themselves. However, we usually think of the mass of an object as a fixed, unmanipulable feature of it – what accounts for the variability in the acceleration of objects is the various forces to which they are subjected. It is at least in part for this reason that (3.1) strikes us an unsatisfactory candidate for a cp law: the factor (or at least the most important factor) – the applied force – that makes a difference to the value of the acceleration is not made explicit in (3.1), but is instead hidden in the completing condition. Indeed, it is just this fact that I drew attention to when I noted that (**F2**) and (**PR**) yield the conclusion that "all masses of 5 kg accelerate at n meters/s^2" are cp laws for *all* values of n. This strikes us as unsatisfactory precisely because we think a serious candidate for a causal generalization should tell us about the conditions which make for variation in the acceleration or under which the acceleration will have one value rather than another, and (3.1) fails to do this.

It is true of course that *some* difference-making factor is left out of most garden-variety causal generalizations including, for example (E). However, (E) does tell us about a variable (treatment) manipulation of which will make a difference to recovery in a wide variety of subjects, given other facts about those subjects. The corresponding claim is not true of (3.1). If we want information about how to change the acceleration of 5 kg. objects, (3.1) by itself without any information about the role of forces, is useless. Similarly for "All humans in Western Europe speak Indo-European languages" – all of the information that has to do with the factors that are relevant to variation in language-speaking ability is again buried in the unknown completing condition.

It is also worth noting that X can make a difference to Y even though there is no condition K that in conjunction with X is nomically sufficient for Y and hence no completer for X with respect to Y. Assume, for the sake of argument, that the action of the chemotherapy drugs on recovery is genuinely indeterministic in the sense that there simply is no additional condition that in conjunction with administration of the drugs is nomically sufficient for recovery. Obviously, this is consistent with its still being true that treatment groups with tumors which receive A systematically have a higher incidence of recovery than control groups which do not receive A or any other anti-tumor agent. Under the assumption of indeterminism, we may still use the randomized design described above to test (E). Nomic sufficiency (or for that matter satisfaction of (**F2**) or (**PR**)) is thus also not necessary for A to make a difference to R. This is another objection to accounts of the generalizations of the special sciences that rely on the notion of a completer – under indeterminism, (E) will be true as long as administration of the drugs make a difference to recovery, even if there is no condition which in conjunction with administration is sufficient either for recovery or non-recovery.

Similar considerations explain the inadequacy of (**M**) or any other requirement that "most Fs be followed by Gs". If the relevant content of (E), from the point of view of explanation and testing, has to do with the fact that whether or not the drug is administered makes a difference for recovery or the probability of recovery, (E) can be true even if only a small portion of those who are treated recover, as long as the incidence of recovery is higher among those who are treated than those who are not. Hence satisfaction of (**M**) is not necessary for (E) to be true. Conversely, consider true claims of the following form:

(W)   Most mammals are or have X.

where X = hearts, hair, mass less than 100kg, non-marsupial etc.

It is plausible that (W) satisfies (**M**) as well as (**F2**) and (**PR**) for many values of X. Nonetheless, I think that there is a clear intuitive sense in which (W) does not by itself describe a condition such that changes in that condition make a difference for or are relevant to whether an animal is X – (W) does not tell us what being X depends on. (This will be particularly plausible if X is also a trait possessed by most non-mammals.) This again shows how proposals like (**F2**), (**PR**), and (**M**) fail to capture the notion of one factors being causally relevant to or making a difference for another.

This points just made are, I think, relatively independent of the details of which particular account we adopt of what it is for one factor to make a difference for or to be causally relevant to another. However in order to have something concrete before us, I propose the following (cf. Woodward 2000, forthcoming): we should think of (E) and other causal generalizations of the sort found in the special sciences as claims about the outcomes of a (possibly hypothetical) series of experiments. Let me begin with a *minimalist* account (see below) of the content of (E) along these lines: (E) claims that for some persons with tumors there is a possible idealized experimental manipulation (a possible intervention) which determines whether they receive A, and that if such interventions were to be carried out on these persons whether they recover (or the probability of their recovering) would be counterfactually dependent on whether or not they receive A. Of course we cannot both give and withhold A from the same individual, at least at the same time, but we can use other experiments, such as the randomized experiment described above, to provide good evidence for the truth or falsity of claims about what whether a population contains individuals for whom recovery is counterfactually dependent on whether they receive A. The notion of administration of treatment being relevant to or making a difference for recovery is thus captured by the notion of the counterfactual dependence of R on A, where the antecedents of the relevant counterfactuals are understood as realized by processes that have the characteristics of idealized experimental interventions. Such an account makes it clear why the sorts of experiments described above are relevant to the truth of (E).

I emphasize that on this minimalist account what is required for the truth of (E) is simply that for *some* individuals, recovery counterfactually depends on some possible intervention that realizes A. It is not required that for *all* people with tumors of type T, recovery counterfactually depends on A. Instead, it is consistent with the truth of (E) that some people have no possibility of recovering even if given A. Nor is it required that all interventions realizing A change whether recovery occurs (or the probability of recovery occurring) for all or even some individuals (see note

4). This account makes it obvious how the content of (E) is very different from the content assigned to it by the various versions of the completer strategy considered above and also makes it clear how in establishing that (E) is true we needn't think of ourselves as establishing a generalization that has the sort of universality and exceptionlessness characteristic of a law of nature

I have called this account of the content of (E) as "mininalist" because in this case, as in the case of many other causal claims, we know a great deal more about the relationship between A and R than merely that As cause Rs. As I understand them, claims of the form (G) Cs cause E are vague or at least non-committal in several important respects. First, they tell us (on the account urged above) merely that *some* change in C will (for some individuals) be accompanied by some change in E or in the probability of E. Of course, one typically would like to know a great deal more about the functional form of the relationship between C and E – ideally, exactly which changes in C will produce which changes in E, or exactly how the probability of E will change under different interventions on C. Or, if this is information is not available, one would at least like to know how changes in the frequency of C in various populations will change the frequency of E. In the example under discussion, researchers have rough information of at least this last sort – they know the frequency of recovery for various sorts of tumors under various chemotherapy regimes for representative samples from the U.S. population, and information about the relative rates of recovery under various treatment regimes – e.g., that certain drugs are more effective than others. Of course, both information about the functional form linking C to E and information about how changes in the frequency of C are linked to changes in the frequency of E also can be given an "interventionist" or "manipulationist" interpretation in an obvious way. Nonetheless such information falls far short of anything that might be plausibly described as a strict or probabilistic law.

An additional respect in which we often know more about the relationship between Cs and Es than that the former cause the latter is this: we have additional information about at least some of the background circumstances in which Cs can or cannot or are particularly likely or unlikely to cause Es. As we shall see below, such information about (what we might call) the *domain* of the generalization linking Cs to Es may be (and often is) vague and informal – sufficiently so, that it may be unsuitable for incorporation into the antecedent of anything that we might wish to call a law. Nonetheless such information can play an important role in explanation and testing.

The account of claims like (E) offered above is thus not meant to deny the obvious point that the more one learns about the conditions under which A will produce recovery for different sorts of individuals, and the more one learns about which realizations of A will produce recovery and which will not, the better. My contention is rather that we should think of establishing more precise generalizations (including, if this turns out to be possible, exceptionless generalizations) about the relationship between chemotherapy and recovery as a limiting case – showing that A causes recovery for some individuals does *not* require that we already be in possession of such more precise generalizations. Instead, the order of discovery is typically just the opposite: we *first* establish that A produces recovery in some cases and then gradually extend what we know to generalizations of greater precision and scope. And, even when we have information about, so to speak, the exportability of causal claims established in one set of background circumstances to new situations, this information often lacks the precision that we demand of laws.[11]

From the point of view of my present discussion what is crucial not whether the interventionist account just described is correct in every respect, but rather the more general point that the claim that F is causally relevant to or of makes a difference to G, however this is understood in detail, is different from and neither entails nor is entailed by the satisfaction of conditions like (**F2**), (**PR**) or (**M**) and that we require some account of what claims like (E) mean that respects this point. Once we recognize this, we should also be able to see that the picture of testing assumed in Section 1 above is far too restrictive. If the only way to test a generalization like (E) is to derive from it and from independently established information about the initial conditions holding for some individual subject (information about whether the subject is treated and other facts about the subject that we can measure such as the genetics of the tumor) a prediction of whether that subject recovers and then check whether that prediction holds, then (E) would indeed be untestable. It is perhaps some thought along these lines that underlies the view about testing adumbrated under (4). But as the experiment described above illustrates, there are other ways of testing (E).

Indeed, randomized experiments are just one of many possible ways of testing and providing evidence for causal generalizations that, like (E), fail to state nomically sufficient conditions for effects. Just confining ourselves to the social and behavioral sciences, books like Cook and Campbell's *Quasi-Experimentation* describe a variety of experimental procedures for establishing causal generalizations that do not involve randomization. Moreover, there is a very large and detailed literature in econometrics, statistics, epidemiology, computer science and philosophy that describes procedures and techniques for reliable inference from non-experimental (observational) data to causal generalizations. These include procedures that rest on relatively domain-general assumptions (e.g. Spirtes, Glymour and Scheines, 2000) as well as those that rely more heavily on domain specific background information. There are also any number of examples of more informal but nonetheless convincing arguments in these disciplines that establish causal conclusions from non-experimental observations.[12] A common thread running through these procedures and arguments is that they often involve either creating new situations or discovering naturally occurring ones in which some or all other possible causes of some effect E besides some putative cause C are known to be distributed on average in the same way across different levels of C.[13] In such cases one may make inferences about the effect of C on E without specifying what these other causes of E are and hence without being in a position to establish any exceptionless generalizations or strict laws about the circumstances under which Cs are always followed by Es. It is baffling that so little attention is paid to the logic of such arguments in the cp laws literature. More generally, it seems to me that if one wants to understand how generalizations that fall short of the standards for strict lawhood can nonetheless be tested and confirmed, it is far more profitable to focus directly on the rich literature on problems of causal inference in the special sciences that now exists, rather than on the problem of providing an account of the truth conditions for *ceteris paribus* laws.

Just as the worries about testing and confirmation that animate the cp laws literature seem to rest on an overly restrictive account of these activities, so also for explanation. One of the main worries that drives the cp literature, evident from the passage quoted from Pietroski and Rey above, is that if the generalization "All Fs are Gs" has exceptions, then we cannot appeal to that generalization and the fact that some object is F to explain why it is G. Clearly this worry presupposes that it is at least a necessary condition for explanation that the explanans be nomically sufficient for the explanandum. I reject this claim. On my view, the core idea in explanation is that an explanans should cite variables or factors that make a difference for the explanandum in the sense described above. As I have shown, for this condition to be satisfied it is neither necessary nor sufficient that the explanans provide a nomically sufficient condition for the explanandum. Thus I would claim that we can explain, by appealing to (E), why the expected rate of recovery of tumor patients in a treatment group which has been administered drugs in accord with A is higher than in a control group that receives no drugs, even though (E) does not state a condition that is nomically sufficient for recovery in individual patients. More controversially, if the right additional conditions are met, we can also appeal to (E) and the fact of treatment in accord with A to explain why an individual patient has recovered.[14] In both cases, these explanations will be genuine because (or to the extent that) they correctly tell us what their explananda (the higher expected incidence of recovery, recovery in the individual patient) depend on (administration of the drugs). Again, once we realize that making a difference is the key notion in explanation, we see that don't need to construe a generalization like (E) as cp law in order to make sense of its role in explanation. Put more negatively, it is not surprising that if we follow the cp literature in ignoring or failing to exploit this additional content of (E) having to do with the idea that changes in A will change or make a difference for R, we will fail to produce an adequate account of generalizations like (E) or how they can be tested or figure in explanations.

I want to conclude this section by commenting briefly on the very different view of the status of claims of claims like (E) adopted by Earman, Roberts, and Smith (*this issue*). Taking as their example (S) "Smoking causes lung cancer", these writers contend that this claim is either "vacuous" or that it should be understood as an "elliptical and imprecise expression" of a much larger body of information (about, e.g., the statistical relationship between smoking and lung cancer) along with a "signal" that indicates commitment to a certain research program. It should be clear from my remarks above that I agree that claims like (S) are non-specific in that they do not tell us exactly which manipulations of whether subjects smoke will change whether they develop lung cancer (or the probability of their developing lung cancer, and if the latter by how much) or in what circumstances. However, it is simply a mistake to take claims like (S) as compressed summaries of a body of statistical information. As a large body of work has made clear,[15] statistical information like "Smoking and lung cancer are correlated" or even "Smoking is correlated with lung cancer across a range of different populations" or "Smoking remains correlated with lung cancer when we control for various other causes of smoking" underdetermines which of several competing claims about the causal relationship between smoking and lung cancer are true. For example, the claims that (i) (S), (ii) lung cancer causes smoking, (iii) smoking and cancer are due to a common cause, as well as any combination of (i), (ii) and (iii), are all consistent with the claim that smoking and lung cancer are correlated. (S) also has additional content over and above the other correlational claims described above. My way of cashing out the difference in content among (i), (ii) and (iii) is in terms of what would happen under interventions – for example, (S) claims that some intervention on whether subjects smoke is associated with changes in whether they develop lung cancer but does not imply that an intervention on lung cancer changes smoking, (ii) claims that some intervention on lung cancer will change smoking and so on. Whether or not this construal of claims like (i)–(iii) is correct, we need some story about what such claims say that does justice to the fact that they are not just summaries of statistical information.

It should also be obvious that on my construal the claim that Cs cause Es is *not* tantamount to the vacuous claim that Es are followed by Cs except when they are not. On my construal, to establish that Cs cause Es (i.e., that some interventions on C will change E) we must have recourse either to controlled experiments or to some body of information that allows us to reach reliable conclusions about what would happen in such an experiment. Claims about what would happen to E under an intervention on C cannot be established simply by observing that Cs are sometimes followed by Es.

4.

As noted above, claims like (E) do not even have the form of exceptionless generalizations and are not functionally precise. I want to conclude this paper by describing briefly how the account just sketched can be extended to more functionally precise generalizations. Consider the claim that a particular kind of neural circuit conforms to the simple Hebbian learning rule

(L) $\Delta w_{kj}(t) = \eta y_k(t) x_j(t)$

where $\Delta w_{kj}(t)$ is the change in synaptic weight of neuron k with presynaptic and postsynaptic signals x_j and y_k and η is a positive constant.

There are at least two respects in which (L) does not fit naturally into the standard (U) "All As are Bs" framework for laws of nature. First, even for those circuits that do conform to (L) there are many possible occurrences, including gross anatomical damage, that would make it the case that those circuits no longer conform to (L) and these occurrences are not delimited by (L) itself. Second, there are many neural circuits that do not conform to (L) but rather conform to some other learning rule. This in itself would not create problems for fitting (L) into the framework (U) if we possessed some way of distinguishing those circuits that conform to (L) from those that do not that was of the appropriate sort for incorporation into the antecedent of a "law". (Appropriate here means that the distinguishing property must be specified in a non-trivial, non-circular way – e.g., that we can't just indicate the circuits that do conform by pointing or by saying that they are often found in the cerebellum and rarely in the cortex.) Let us suppose, however, as seems likely, that we do not know of such a property. Do we need to reinterpret (L) as a cp law and then search for conditions under which its cp clause is non-vacuous in order to vindicate its scientific status?

Just as with (E) it seems to me that we do not. The natural way of understanding the claim that some particular kind of neural circuit conforms to (L) is in interventionist terms: for such neurons and in some appropriate background circumstances some interventions that change $x_j(t)$ will change $\Delta w_{kj}(t)$ in accord with (L). One can establish that this claim is true (by for example, experimentation) for some particular kind of circuit (e.g., in the hippocampus) without being able to enumerate all of the circumstances in which (L) breaks down and without being able to specify the circumstances in which it holds in a way appropriate for inclusion within the antecedent of a law. For example, even if it is true that (L) would break down under certain kinds of anatomical damage (and we are currently unable to specify in an informative, non-circular way what those circumstances are) we can nonetheless test and confirm or disconfirm whether, given those background conditions that do in fact occur (which, let us suppose, in fact include no such damage), a particular circuit or kind of circuit conforms to (L). No special account of the semantics of the cp clause or its role in testing is required to explain how to do this.

The temptation to suppose otherwise derives, as I see it, from the assumption that if (L) is "scientifically legitimate" it must be possible to formulate it as an exceptionless generalization of form

(L*)   For all neural circuits with property P, then $\Delta w_{kj}(t) = \eta y_k(t) x_j(t)$

Given this assumption, it is indeed a serious – in fact, insoluble – problem that we are unable to specify P in an informative, non-trivial way. But, as I have sought to show, in order to use (L) to explain, model, or predict the behavior of individual circuits, we simply don't need to formulate a generalization along the lines of (L*). The (at least partly unknown) conditions P that must be satisfied in order for individual circuits to conform to (L) should not be thought of as part of the antecedent of some "law" involving (L) but rather as conditions for the applicability for (L). And as I have argued elsewhere (Woodward 2000), it is not always appropriate to think of the conditions for the applicability of a generalization – at least as we are able to formulate them – as candidates for incorporation into the antecedent of that generalization. In the case of many generalizations of the special sciences, including, one suspects, (L), we may have no alternative to either learning "empirically" or "inductively" on a case-by-case basis (that is, in a way that does not yield some interesting theoretical characterizations of what is common to all such cases) whether the conditions of applicability for the generalization are satisfied or else to specifying those conditions in a loose, informal, or ad hoc way that makes them unsuitable for incorporation into the antecedent of a law.[16] If the argument of this paper is correct, there need be nothing wrong with this way of proceeding and it need not result in generalizations that are vacuous or untestable.

ACKNOWLEDGMENTS

Thanks to Alan Hajek, Chris Hitchcock, and Jesse Prinz for helpful comments and discussion.

NOTES

[1] The literature on cp laws bears witness to this point. Although the standard strategy is to characterize the notion of a cp law by reference to the notion of an unqualified or strict (exceptionless) law, the latter notion is usually left undefined, aside perhaps from a few hand-waving references to "standard criteria for lawfulness" – criteria which most philosophers agree are inadequate for distinguishing laws from non-laws (see van Fraassen, 1989, Woodward, forthcoming). It is curious that the inability of anyone to say clearly what a law of nature is has not had more of an impact on the cp laws industry.

[2] If the gravitational inverse square law is understood as describing the component of the gravitational force experienced by a mass that is due to some second mass, then there is no need to think of it as having exceptions whenever non-gravitational forces are present and this particular motivation for construing it as a *ceteris paribus* law is removed.

[3] After completing a draft of this paper, I became aware that very similar conclusions about the relationship between the strategy under discussion and determinism are reached in Schurz, 2001. Among other things, Schurz proves that given determinism and very weak additional assumptions, it follows from the characterization of *ceteris paribus* laws in Pietroski and Rey (1995) that "*Ceteris paribus*, all As are Cs" is a *ceteris paribus* law for arbitrary A and C – a result which, as Schurz says, shows that Pietroski's and Rey's characterization is "almost vacuous" (p. 359). I thank John Earman for drawing my attention to Schurz's paper. A similar conclusion is also reached in Woodward 2000.

[4] What Fodor actually requires under his Story 2 is that "All Fs are Gs" is a cp law if and only if every realization of F has a completer and it is nomologically possible that that a token of F occurs without the occurrence of a token of its completer (1991, p. 24). This is too permissive in the sense that it is subject to the counterexamples to (**F2**) described below. In addition it seems to me that the demand that every realization of A have a completer is too stringent in the sense that it is not a necessary condition for the truth of claims like (E), at least if biomedical practice is our guide. It is not at all far fetched to suppose that some realizations of A in our example will not be effective against any form of the tumor – e.g., perhaps therapies delivered between fourteen and twenty days apart, although consistent with the protocol and hence instances of A, are always ineffective, although this fact is

unknown at present to researchers. As I argue below, all that is required for the truth of (E) is something like this: *some* instances of A cause recovery in the actual circumstances. It is not required that for every instance of A there be some circumstance in which A is *always* followed by recovery.

In his (1991) Fodor goes on to offer a second account of cp laws in terms of "networks" of laws. For convincing criticism of this account see Mott, 1992.

[5] In addition to the problems described below, there are other more subtle difficulties with both approaches, centering on the question of whether it is plausible to suppose that the conditions (i)–(iii) for the existence of a completer will be simultaneously satisfied in the case of all true causal generalizations. Consider (E). As noted above, the property or variable A is really a lengthy disjunction of more specific properties – $A_1 v\ A_2 v\ \ldots A_n$ – that differ in the causal effects. Similarly, the candidate completing condition C for (E) is also a complex disjunction – $C_1 v\ C_2 v\ \ldots\ C_m$ – where each C_i captures causally relevant features of the patient and tumor that vary from patient to patient. Assuming that underlying processes are always deterministic, what is really going on when (E) is true is that some pairs $A_i C_j$ always lead to recovery and other pairs do not, although there may be no A_i that leads to recovery for all C_j and no C_j such that recovery follows for all A_i. The question we are then faced with is whether we can specify a K such that (E*) "All As in K are Rs" is a law and the conditions (ii)–(iii) for Ks being a completer are satisfied. It is hard to see that the truth of (E) *guarantees* that this is possible. For example, one way of turning (E) into an exceptionless generalization of form (E*) would be to add to its antecedent some condition K* that is equivalent to the disjunction of those $A_i C_j$ that are always followed by R. However, K* by itself is sufficient for R in violation of (iii). What reason do we have for thinking that a K* satisfying (i)–(iii) will always exist? I will ignore such complications in the rest of my discussion.

[6] See Woodward, 2000 for additional discussion of this example.

[7] This way of looking at matters contrasts with the position taken in Earman, Roberts, and Smith (*this volume*). These writers seem to suggest that there is no interesting in principle problem with the use of cp laws if the cp clause they contain is eliminable in favor of a precise condition. I disagree. As I see it the central problem raised by cp laws or any other set of special science generalizations is to explain how those generalizations can be tested and can figure in explanations given what we now know and given the content we presently attribute those generalizations. For example, what we need to explain is how (E) just as it stands can be tested and can figure in explanations. It is not an adequate treatment of this problem to note that if (E) were replaced with a strict law, that law could be tested. For similar reasons, even if it is true (as I believe that it is) that underlying (E) is some exceptionless generalization, this fact does not by itself explain how (E) can be tested or what evidence is relevant to confirming it.

[8] Again, we might replace (V) with (V**) "All Fs are Gs except when they are not and furthermore most Fs are Gs" or stipulate that (V**) is to be regarded as part of the content of (V). (V**) is not vacuous, but only because it is a round about way of saying that most Fs are Gs.

[9] It may be tempting to think that the suggestion that (E) may be tested by means of a randomized experiment amounts to a vindication of the claim that (E) is a *ceteris paribus* law. After all, in testing (E) in this way, do we not create a situation in which "other things are equal" in the sense that (on average) other factors that affect recovery besides A are distributed in the same way across the treatment and control groups? This is a mistake. The fact that one way of testing (E) is to create treatment and control groups across which

"other things are equal" does not show that a *ceteris paribus* clause is part of the content of (E) or that (E) holds only when "other things are equal". For one thing, although we may test (E) by means of a randomized experiment, we expect (E) to hold across a great many other contexts, including those in which no randomization is present.

[10] I thank Jesse Prinz for forcefully drawing my attention to this line of argument.

[11] Illustrations of this general point are provided in Woodward, 1995. I argue there that a typical pattern in the social sciences is that numerically precise causal generalizations are population-specific. What generalizes across different populations is more qualitative information about capacities, mechanisms, and the direction or broad magnitude of effects. For example, the series of experiments carried out in different American states in the 1970s to test the effects of a negative income tax on labor market participation found a broadly similarly qualitative pattern in all of the populations studied: the negative income tax has a relatively small effect on the labor-market response of primary wage earners and a considerably more substantial effect on secondary wage earnings. Nonetheless, the coefficients representing the quantitative values of these effects vary in nontrivial ways across these different populations studied. Estimates of uncompensated wage elasticity for the entire population of U.S. adult males range from -0.19 to -0.07 and estimates of income elasticity range from -0.29 to 0.17. Estimates of uncompensated wage elasticity from two negative income experiments (Gary, and Seattle-Denver) yield values of 0 (Gary), -0.06 (short-run, Seattle-Denver), and 0.02 (long-run, Seattle-Denver) (Stafford, 1985). The qualitative generalization that the negative income tax has a relatively small effect on the labor-market response of primary wage earners and a more substantial effect on secondary wage earning represents genuine causal information that holds across a range of populations and background circumstances. It is a good example of a non-vacuous causal generalization that falls well short of being a "strict" law.

[12] In epidemiology, John Snow's discovery (Snow 1855) of the causes of cholera is a classic example.

[13] The familiar assumption in the context of multivariate regression that the error term is uncorrelated with any of the independent variables is one illustration of this.

[14] The simplest case in which we may appeal to (E) to explain individual outcomes is one in which our background knowledge tells us that there is no possibility of recovery in the absence of some form of chemotherapy and a patient receives A and recovers. In this sort of case, A explains the recovery in the sense that A is what the recovery depends on or what made the difference for recovery.

[15] Important results about the extent of the underdetermination of causal claims by one kind of statistical information – information about statistical independence relationships – can be found in Spirtes, Glymour and Scheines, 2000. I should acknowledge in this connection the important point that there are special cases in which independence relationships uniquely determine the causal facts, given certain additional principles (what these authors call the Causal Markov Condition and Faithfulness). However, no such uniqueness result holds in general.

[16] A somewhat artificial but (I hope) straightforward illustration: We have a population of springs, some of which conform (within a certain range of extensions) to a particular specification of Hooke's law (H) $F = -k_s X$, and some of which do not. The only way we can pick out the springs conforming to (H) is by extending them and measuring the force they exert, although having done this we can then mark them off in some way (e.g., by assigning them a number) or else noticing some property – e.g., the presence of a red mark – that they happen to have in common. Having identified both groups we can then retest

to determine whether those springs continue to conform to (H) or not and under which circumstances.

Note that while this procedure yields substantive non-tautologous causal information, we have not formulated any exceptionless "law" of the form

(S)   If y is a spring and Cy, then the force exerted by y when extended conforms to (H).

In particular the property (the numbers, the red mark) used to pick out the springs conforming to (H) is not an acceptable candidate for C. It is *not* a matter of law that all the springs that are numbered or red conform to (H): we cannot make it the case that some spring that previously did not conform to (H) now does conform by marking it red or numbering it. This illustrates the general point that the conditions we employ for applying a causal generalization may not be suitable for incorporation into the generalization itself.

REFERENCES

Cartwright, N.: 1983, *How the Laws of Physics Lie*, Clarendon Press, Oxford.

Cook, T. and D. Campbell: 1979, *Quasi-Experimentation: Design and Analysis Issues for Field Settings*, Houghton Miflin, Boston.

Earman, J. and J. Roberts: 1999, '*Ceteris Paribus*, There Is Problem of Provisos', *Synthese* **118**, 439–478.

Earman, J., J. Roberts, and S, Smith: This volume, '*Ceteris Paribus* Lost', *Erkenntnis* **57**, 281–301.

Fodor, J.: 1991, 'You Can Fool Some of the People All of the Time, Everything Else Being Equal; Hedged Laws and Pscychological Explanation', *Mind* **100**, 19–34.

Glymour, C.: This volume, 'A Semantics and Methodology for *Ceteris Paribus* Hypotheses', *Erkenntnis* **57**, 395–405.

Hausman, D.: 1992, *The Inexact and Seperate Science of Economics*, Cambridge University Press, Cambridge.

Mott, P.: 1992, 'Fodor and Ceteris *Ceteris Paribus*', *Mind* **101**, 3350-346.

Pietroski, P. and G. Rey: 1995, 'When Other Things Aren't Equal: Saving *Ceteris Paribus* Laws from Vacuity', *The British Journal for the Philosophy of Science* **46**, 81–110.

Reddy, A., A. Janss, P. Philips, H. Weiss, and R. Packer: 2000, 'Outcome for Children with Supratentorial Neuroectodermal Tumors Treated with Surgery, Radiation, and Chemotherapy', *Cancer* **88**(9), 2189–2193.

Schurz, G.: 2001, 'Pietroski and Rey on *Ceteris Paribus* Laws', *The British Journal for the Philosophy of Science* **52**, 359–370.

Snow, J.: 1855, *On the Mode of Communication of Cholera*, John Churchill, London.

Stafford, F.: 1985, 'Income Maintenance Policy and Work Effort: Learning from Experiments and Labor Market Studies', in J. Hausman and D. Wise (eds), *Social Experimentation*, University of Chicago Press, Chicago.

van Fraassen, B. C.: 1989, *Laws and Symmetry*, Clarendon Press, Oxford.

Woodward, J.: 1995, 'Causality and Explanation in Econometrics', in Daniel Little (ed.), *On the Reliability of Economic Models: Essays in the Philosophy of Economics*, Kluwer, Dordrecht, pp. 9–61.

Woodward, J.: 2000, 'Explanation and Invariance in the Special Sciences', *The British Journal for the Philosophy of Science* **2000**, 197–254.

Woodward, J.: Forthcoming, *Making Things Happen: A Theory of Causal Explanation*, Oxford University Press, Oxford.

James Woodward
Division of Humanities and Social Sciences, 101-40
California Institute of Technology
Pasadena CA 91125-001
U.S.A.
E-mail: jfw@hss.caltech.edu

SANDRA D. MITCHELL

CETERIS PARIBUS – AN INADEQUATE REPRESENTATION FOR BIOLOGICAL CONTINGENCY

ABSTRACT. It has been claimed that *ceteris paribus* laws, rather than strict laws are the proper aim of the special sciences. This is so because the causal regularities found in these domains are exception-ridden, being contingent on the presence of the appropriate conditions and the absence of interfering factors. I argue that the *ceteris paribus* strategy obscures rather than illuminates the important similarities and differences between representations of causal regularities in the exact and inexact sciences. In particular, a detailed account of the types and degrees of contingency found in the domain of biology permits a more adequate understanding of the relations among the sciences.

1. *CETERIS PARIBUS* LAWS AND BIOLOGICAL CONTINGENCY

Biological systems are evolved, multi-component, multi-level complex systems. Their features are, in large part, historically contingent. Their behavior is the result of the interaction of many component parts that populate various levels of organization from gene to cell to organ to organism to social group. It is my view that the complexity of the systems studied by biology and other sciences has implications for the pursuit and representation of scientific knowledge about such systems. I will argue that a proper understanding of the regularities in biological systems should influence our philosophical views on the nature of causal laws and, in particular, the role of *ceteris paribus* qualifications.

A well-known problem for the special sciences, and biology in particular, is the failure of generalizations about evolved, complex systems to meet what have been identified as the defining characteristics of scientific laws. This is alleged to be a serious problem because of the special role that laws play in science. They are what science supposedly seeks to discover. They are supposed to be the codifications of knowledge about the world that enable us to explain why what happens, happens, to predict what will happen in the future or in other circumstances and provide us the tools to intervene in the world in order to reach our pragmatic goals. As such, they have been taken to be the gold standard of modern scientific practice. Philosophers have analyzed and re-analyzed the concept of a scientific

 Erkenntnis **57**: 329–350, 2002.
© 2002 *Kluwer Academic Publishers. Printed in the Netherlands.*

law or a law of nature in the hopes of specifying a set of necessary and sufficient conditions that postulations of laws have to meet in order to be the "real thing" and hence be able to perform the functions of explanation, prediction, and intervention. The "received view" of what conditions are required of a law include:

1. logical contingency (having empirical content)
2. universality (covering all space and time)
3. truth (being exceptionless)
4. natural necessity (not being accidental)

Some hold that laws are not just records of what happens in the universe but are stronger claims about what must happen, albeit not logically, but physically in our world and hence have the power to dictate what will happen or what would have happened in circumstances which we have not in fact encountered. Thus laws are said to support counterfactuals. It is not clear that anything that has been discovered in science meets the strictest requirements for being a law. However, if true, presumably Newton's Laws of Motion, or The Laws of Thermodynamics, or the Law of the Conservation of Mass/Energy, would count. The closest candidates for being a law and test cases for a philosophical account of scientific law live most commonly and comfortably in the realm of physics. Many philosophers have pointed out the fact that few regularities in biology seem to meet the criteria for lawfulness enjoyed by the laws of physics.

How are we to think about the knowledge we have of biological systems that fail to be characterized in terms of universal, exceptionless, necessary truths? Their inferior status is sometimes blamed on the contingency of biological causal structures. The ways in which biological systems are organized has changed over time, they have evolved. Their causal structures thus not only could have been different but in fact were different in diverse periods in the evolution of life on the planet and in distinct regions of the earth and most likely will be different in the future. Thus exceptionless universality seems to be unattainable. The traditional account of scientific laws is out of reach for biology. Should we conclude that biology is lawless?

If so, how can we make sense of the fact that the patterns of behavior we see in a social insect colony or the patterns of genetic frequencies we see over time in a population subject to selection are caused, are predictable, are explainable, and can be used to reliably manipulate biological systems? The short answer is that biology has causal knowledge that performs the same epistemological and pragmatic tasks as strict laws without being universal, exceptionless truths, even though biological knowledge consists of contingent, domain restricted truths. This alone raises the question of whether laws in the traditional sense should be taken as the gold standard against which to assess the success or failure of our attainment of scientific knowledge.

But perhaps we should not be too quick to abandon the standard. There is, after all, a well-worn strategy for converting domain restricted, exception ridden claims into universal truths and that is by means of the addition of a *ceteris paribus* clause. Take the causal dependency described by Mendel's law of segregation. That law says; in all sexually reproducing organisms, during gamete formation each member of an allelic pair separates from the other member to form the genetic constitution of an individual gamete. So, there is a 50:50 ratio of alleles in the mass of the gametes. In fact, Mendel's law does not hold universally. We know two unruly facts about this causal structure. First, this rule applied only after the evolution of sexually reproducing organisms, an evolutionary event that, in some sense, need not have occurred. Second, some sexually reproducing organisms don't follow the rule because they experience meiotic drive, whereby gamete production is skewed to generate more of one allele of the pair during meiotic division. Does this mean than Mendel's law of segregation is not a "law"? We can say that, *ceteris paribus*, Mendel's law holds. We can begin to spell out the *ceteris paribus* clause: provided that a system of sexual reproduction obtains, and meiotic drive does not occur, and other factors don't disrupt the mechanisms whereby gametes are produced, then gamete production will be fifty-fifty. Finer specifications about possible interference, *especially when they are not yet* identified, get lumped into a single phrase – "*ceteris paribus*" – when all else is equal, or provided nothing interferes. This logical maneuver can transform the strictly false universal claim of Mendel's law into a universally true, *ceteris paribus* law. With the *ceteris paribus* clause tacked on, even biological generalizations have the logical appearance of laws.

But, the cost of the *ceteris paribus* clause is high. First, although making a generalization universally true in this way can always be done, it is at the risk of vacuity. Woodward (this volume) makes this argument clearly and rejects *ceteris paribus* laws entirely, advocating instead a revision of our account of explanation that does not require universality. Others, like Pietroski and Rey (1995) have suggested that there are ways to fill out the *ceteris paribus* clause to make it contentful. However, the ability to fully fill in the conditions that could possibly interfere may well be an impossible task. Indeed, in evolutionary systems new structures accompanied by new rules may appear in the future, and hence we could never fully specify the content of potential interfering factors. Still others, Lange (2000, this volume) have argued that vagueness is not equivalent to vacuity.

He argues that scientists in their practice tacitly know what is meant by a *ceteris paribus* law. They know some cases of interfering factors and can extrapolate the nature of other factors by means of their family resemblance to the known ones. Earman, Roberts, and Smith (this volume) maintain that there are strict laws to be found, at least for fundamental physics, so there is no need for *ceteris paribus* laws there. Furthermore, they argue, that although the special sciences cannot discover strict laws, there are no such things as *ceteris paribus* laws. Their challenge leaves us with the problem of how to account for the explanatory and predictive power of biological generalizations if, as their account would entail, there are no laws in these domains. I will argue that it is only by providing a detailed account of biological knowledge claims that we can hope to address the problem that Earman, Roberts and Smith have posed.

Critics of the *ceteris paribus* clause correctly identify the fact that the clause violates the logical spirit of the concept of 'law'. I will argue that, more importantly, it violates a pragmatic aspect of 'laws' in that it collapses together interacting conditions of very different kinds. The *logical* cloak of *ceteris paribus* hides important differences in the *ontology* of different systems and the subsequent differences in epistemological practices. Where as *ceteris paribus* is a component of the statement of a causal regularity, what it is intended to mark in the world is the *contingency* of the causal regularity on the presence and/or absence of features upon which the operation of the regularity depends. Those contingencies are as important to good science as are the regularities that can be abstracted from distributions of their contextualized applications.

Indeed, a familiar way to mark the difference between exact and special sciences is by pointing to the contingency of the products of biological evolution, and the contingency of the causal dependencies to which they are subject. This is what Beatty dubs 'the evolutionary contingency thesis' (Beatty 1997). It is meant to capture the meaning of Steven J. Gould's metaphoric appeal to Frank Capra's "It's a Wonderful Life" (Gould 1990). That is, if we rewound the history of life and 'played the tape' again, the species, body plans and phenotypes that would evolve could be entirely different. The intuition is that small changes in initial 'chance' conditions can have dramatic consequences downstream. Sexual reproduction itself is thought to be a historically contingent development and hence the causal rules that govern gamete formation, for example, are themselves dependent on the contingent fact that the structures that obey those rules evolved in the first place. Biological contingency denotes the historical chanciness of evolved systems, the 'frozen accidents' that populate our planet, the lack of necessity about it all.

There have been different responses to the mismatch between the strict, ideal version of scientific law and the products of the special sciences.

- Biology has NO LAWS on the standard account of laws (Beatty, 1995, 1997, Brandon)

    ○ This may not be so bad, if as Cartwright argues science doesn't strict need laws or as Woodward argues we can have explanatory generalizations without universality.

- Biology HAS LAWS on the standard account of laws (Sober, Waters)

    ○ There are not many and these are *ceteris paribus* laws or are very abstract, perhaps mathematical truths or laws of physics and chemistry.

- Biology HAS LAWS on a revised account

    ○ We need to reject the standard account of laws and replace it with a better account (Mitchell)

Some have opted to accept the lesser status of biological generalizations and preserve the language of law for those venerable truths that people like Steven Weinberg dream will be few and so powerful as to make all the other knowledge claims we currently depend on part of their deductive closure. If laws are understood in the strict sense, biology doesn't have any, and the picture is even worse for the social sciences (Beatty 1995, 1997; Brandon 1997). Others while accepting this conclusion have gone on to suggest that laws aren't all they are cracked up to be, anyway, and so it is not so bad that the special sciences fail to have them – we don't need them (Cartwright 1994, 1999). Still others scramble to construe the most abstract of relationships within the special sciences or some physical or chemical regularity internal to biological systems as laws (Sober 1997; Waters 1998), thus there are some, though not many and not clearly identified as distinctively biological (Brown and West 2000). There are some, nevertheless, and hence the legitimacy of the special sciences can be restored according to this line of argumentation.

I find these responses less than satisfying. We need to rethink the idea of a scientific law pragmatically or functionally, that is, in terms of what scientific laws let us do rather than in terms of some ideal of a law by which to judge the inadequacies of the more common (and very useful) truths. Woodward (2002, this volume) also adopts a strategy to reconsider the nature of laws in the special sciences, rather than forcing those claims uncomfortably into the standard view, wedged in with the help of *ceteris paribus* clauses. He had developed an account of explanation that requires

SANDRA D. MITCHELL

SANDRA D. MITCHELL

generalizations less than universal in scope, but which can, nevertheless, support counterfactuals.

My strategy is somewhat different: I recommend that we look more closely at the character of the contingencies of the causal dependencies in biological systems that are often lumped into a single abstract concept of 'contingency' and singled out as *the* culprit preventing biological science to be lawful (Beatty 1997). I have argued elsewhere (1997, 2000) that general truths we discover about the world vary with respect to their degree of contingency on the conditions upon which the relationships described depend. Indeed, it is true that most of the laws of fundamental physics are more generally applicable, i.e., are more stable over changes in context, in space and time, than are the causal relations we discover hold in the biological world. They are closer to the ideal causal connections that we chose to call 'laws'. Yet, few of even these can escape the need for the *ceterus paribus* clause to render them logically true. Indeed, Cartwright has argued that we can't find a single instantiation of these laws and Earman et al. in this volume admit that we have yet to identify (or at least have evidence that we have identified) a law of physics. What's going on here?

The difference between fundamental physics and the special sciences is *not* between a domain of laws and a domain of no laws. Yet, I would agree, there are differences and those differences can inform our understanding of not only the special sciences but of the very notion of a 'law' and its function. By broadening the conceptual space in which we can locate the truths discovered in various scientific pursuits we can better represent the nature of the actual differences. The interesting issue for biological knowledge is not so much whether it is or isn't just like knowledge of fundamental physics, but how to characterize the types of contingent, complex causal dependencies found in that domain.

Recent developments in understanding scientific causal claims take as their starting point less than ideal knowledge. They explore how to draw causal inferences from statistical data, as well as how to determine the extent of our knowledge and range of application from the practices of experimentation. (Spirtes et al., 1993; Glymour 2001; Woodward 2000, 2001, 2002). When we see that some relationship holds here and now for this population in this environment, what can we say about the next population in the next environment? These new approaches recognize the less-than-universal character of many causal structures. If explanatory generalizations were universal, then when we detect a system in which A is correlated with B, and we determine that this is a causal relationship, we could infer that A would cause B in every system. But we often cannot. The difficulty goes beyond the correlation-causation relationship. Even when we gave good evidence that *A* causes *B* in a system, say through controlled intervention, we still can't say that *A* would cause *B* in every system. When we look at the tidy behavior of Mendel's pea plants, where internal genetic factors assort independently and segregate fairly, we might wish to infer that would always happen, in any sexually reproducing population. But it doesn't. And since it doesn't, we need to understand more about the system of Mendel's peas and their relationship to other systems to know what about the original test case is exportable to the new domains. If we were so lucky as to have detected a universal exceptionless relationship, constitutive of the strict interpretation of law, we would know it would automatically apply to all times and all places. But that is not the world of the special sciences.

In systems that depend on specific configurations of events and properties which may not obtain elsewhere, and which include the interaction of multiple, weak causes rather than the domination of a single, determining force, what laws we can garner will have to have accompanying them much more information if we are to use that knowledge in new contexts. These are precisely the domains that the special sciences take as their objects of study. Thus the central problem of laws in the special sciences, and perhaps for all sciences, is shifted from what is a strict law, *ceteris paribus*, or no law at all, to how do we detect and describe the causal structure of complex, highly contingent, interactive systems and how do we export that knowledge to other similar systems (Mitchell 2002).

Let me then consider the parts of this situation in turn. First, what kinds of complexity are present in biological systems? Second, what is that nature of the contingency of causal structures in biology? Third, what are the implications of complexity and contingency for scientific investigation?

2. COMPLEXITY

While the term 'complexity' is widely invoked, what is meant by it varies enormously. Often linked with chaos and emergence, current definitions of complexity numbered somewhere between 30 and 45 in 1996 (at least according to Horgan's report of Yorke's list. See Horgan (1997, fn. 11, p. 197)). I believe the multiplicity of definitions of 'complexity' in biology reflects the fact that biological systems are complex in a variety of ways.

● They display complexity of _structure_, the whole being formed of numerous parts in non-random organization (Wimsatt 1986; Simon 1981).

- They are complex in the *processes* by which they develop from single celled origins to multicellular adults (Raff 1996; Goodwin 1994; Goodwin and Saunders 1992) and by which they evolve from single celled ancestors to multicellular descendants (Buss 1987; Bonner 1988; Salthe 1993).
- The *domain* of alternative evolutionary solutions to adaptive problems defines a third form of complexity. This consists in the wide diversity of forms of life that have evolved despite facing similar adaptive challenges. (Beatty 1995; Mitchell 1997, 2002).

2.1. *Compositional Complexity*

Minimally, complex systems can be identified in contrast to simple objects by the feature of having parts. Simon defined a complex system as "made up of a large number of parts that interact in a nonsimple way ... the whole is more that the sum of the parts" (1981 p. 86). Complex systems are also characterized by the ways in which the parts are arranged, i.e., the relations in which the components stand or their structure. The cells constituting a multicellular organism differentiate into cell types and growth fields (Raff 1996). A honeybee colony has a queen, and workers specialized into nursing, food storage, guarding or foraging tasks (Winston 1987; Wilson 1971). Such systems are bounded, have parts, and those parts differentially interact. In hierarchical organization the processes occurring at a specific level – for example the level of an individual worker in a honey bee colony – may be constrained both from below by means of its genetic make up or hormonal state and from above by means of the demographic features of the colony. In Simon's terms, the system is partially decomposable. Interactions occur in modules and do not spread across all parts constituting the system. For example, in a honey bee colony if a need develops for more pollen, the workers whose task it is to forage for pollen will increase their activity, but changes need not occur in all other task groups (Page and Mitchell 1991).

Different types of organizational structures will mediate the causal relations within a level and between levels. Some modular structures shield the internal operations of a system from external influences, or at least from some set of them. Other features may structure the way in which external information is transmitted through the module.

2.2. *Complex Dynamics*

The non-linearity of mathematical models which represent temporal and spatial processes has become, for some, the exclusive definition of complexity. While this aspect of complexity is widespread it still captures only one aspect of process complexity. Process complexity is linked with a number of dynamical properties including extreme sensitivity to initial conditions (the butterfly effect) and self-organizing and recursive patterning (thermal convection patterns) (Nicolis and Prigogine 1989). Self-organization refers to processes by which global order emerges from the simple interactions of component parts in the absence of a pre-programmed blueprint.

The sensitivity of complex behaviors is further complicated by the fact that developing and evolving organisms and social groups are subject to the operation of multiple, weak, nonadditive forces. For example, in an evolving population natural selection may be operating simultaneously at different levels of organization – on gametes, on individuals with variant fitness relative to their shared environment, on kin groups of different degrees of relatedness, etc. (Brandon 1982; Sober 1984; Hull 1989). In addition to multiple levels of selection, genetic drift may influence the patterns of change in the frequencies of genes through time. Phylogenetic constraints which limit the options for adaptive response, as well as physical constraints, such as the thermal regulation differentials for land dwelling or sea dwelling endotherms, may also stamp their character on the types of change available in an evolving population. One, some, or all of these different forces may operate in varying combinations.

2.3. *Evolved Diversity of Populations*

The third sense of complexity found in biology is exhibited by the diversity of organisms resulting from historical contingencies. Given the irreversible nature of the processes of evolution, the randomness with which mutations arise relative to those processes, and the modularity by which complex organisms are built from simpler ones, there exists in nature a multitude of ways to 'solve' the problems of survival and reproduction. For example, a social insect colony adjusts the proportions of workers active in particular tasks in correspondence to both internal and external factors. If there is a destruction of foragers, the younger individuals may leave their nursing or food storing tasks to fill the vacant jobs. This homeostatic response is accomplished in different ways by different species. Honeybee colonies harbor sufficient genetic variability among the workers to generate variant responses to stimuli (Page and Metcalf 1982; Calderone and Page 1992). Ant colonies, on the other hand, may accomplish the same sort of response flexibility by means not of genetic variability by the variations in nest architecture (Tofts and Franks 1992). The ways in which information gets modulated through these systems depends on a host of properties. But what

is important to notice here is that the manner by which one social insect
colony solves a problem may well be different from solutions found by
similar organisms.

Historical contingencies contribute to the particular ways in which
organisms develop and evolve. History fashions what mutational raw materials are available for selection to act upon; what resources already
present can be co-opted for new functions; and what structures constrain
evolutionary developments.

Complexity carries with it challenges to scientific investigation of
causal dependence and the discovery of explanatory laws. For example,
redundancy of systems to generate a functional state will make experimentation in the standard sense less definitive. What could we learn from
a controlled experiment? Imagine we are trying to determine the consequences of a particular component, say of visual information on the
direction of flight of honeybee foragers. We create two study populations
that are identical for genetics, age, environment, food source, etc. We block
the visual uptake of the individuals in the first population while leaving
the system operative in the second population. Then we look to see what
differences there are in the foraging behavior, ready to attribute differences
to the role of visual information. It could well be that the foraging behavior
in the two populations is identical. Does that tell us that visual information
plays no causal role in foraging decisions? No. It is plausible to postulate
that when visual information is not available, an olfactory system takes
over and chemical cues in that mechanism generate the same behavioral
responses – i.e., ones that are adaptive or optimal responses to foraging
problems.

Redundancy of mechanism makes controlled experimental approaches
problematic. This is not to say that each modality or mechanism cannot be
studied to determine the contribution of each in the absence of the activ-
ation of the others. However, redundancy can take a number of different
forms that make the ways in which the mechanisms mutually contribute to
an overall outcome vary. For example as described in the hypothetical case
above, there may be serial and independent redundant 'back-up' systems.
There may be mutually enhancing or amplifying systems or, conversely,
mutually dampening systems. Knowing the causal laws that govern each
process in isolation by experimentation will not automatically yield sufficient information for drawing an inference to their integrated contribution
in situ. As I will discuss below, this type of complexity introduces a special
set of contingencies in which to consider the operation of any partially
isolatable mechanism.

Another feature of complexity also affects epistemological practices. Consider situations of what we might call phase changes or nonlinear processes whereby the causal relationship governing the behavior of two variables changes completely at certain values of one of the variables. Consider the slime mold, dichtystelium, which lives much of its life as a collection of individual single cells moving through space in search of food. The movements of the cells are driven by the detection of food. However, should a group of cells find itself in a situation in which the value of the food variable is below a threshold, i.e., what would be close to starvation rations, then an entirely new set of causal dependencies kick in. Now instead of each cell moving toward food in a predictable way, the individual cells are drawn to each other, amass together and form a new organism, a multi-cellular slime mold made up of stalk and fruiting body. The latter emits spores to search for more nutrient friendly hunting grounds. The rules governing cellular behavior have changed.

Which rules apply, and when the rules apply is contingent. Complexity affords different kinds of contingency that must be understood to accommodate both the knowledge we have of complex systems and the practices scientists engage in to acquire this knowledge.

3. CONTINGENCY

It is not particularly useful to say *that* laws are contingent or that they can be re-written as *ceteris paribus* generalizations, without detailing *what* kinds of conditions they depend upon and *how* that dependency works. Only by further articulating the differences rather than covering them over with a phrase denoting the existence of restrictions, can the nature of complex systems be taken seriously. The problem of laws in the special sciences is not just a feature of our epistemological failings; it is a function of the nature of complexity displayed by the objects studied by the special sciences. Providing a more adequate understanding of laws in the special sciences requires a better taxonomy of contingency so that we can articulate the several ways in which laws are *not* 'universal and exceptionless'. In what follows I will detail a taxonomy of different sorts of contingencies that can play a role in the operation of a causal mechanism. Knowing when and how causal processes depend on features of what we relegate to the context of their operation is central to using our understanding of casual dependence to explain, predict, and intervene. I have grouped the types of contingency by consideration of logic, spatio/temporal range, evolution and complexity.

3.1. *Logical Contingency*

I have pointed out elsewhere (Mitchell 2000) the obvious point that there is a clear sense in which all truths about our world that are not logical or mathematical truths are contingent on the way our universe arose and evolved. Thus there is one type of contingency, i.e., logical contingency that applies to all scientific claims. The causal structure of our world is not logically necessary. This is as true for physical structures that might have been fixed in the first three minutes after the big bang, atomic structures that appeared as the elements were created in the evolution of stars, and the dependencies found in complex, evolved biological structures whose rules changed from self-replicating molecules with the subsequent evolution of single celled existence, multi-cellularity, sexual reproduction and social groups. All causal dependencies are contingent on some set of conditions that occur in this world, not in all possible worlds.

3.2. *Space/Time Regional Contingency (No Sexually Reproducing Organisms, No Mendelian Law)*

There is another sense of contingency that is attributed to biological laws to distinguish them from the laws of physics which refers to restrictions in spatial and temporal distribution (see also Waters 1998). Mendel's law of segregation of gametes, for example, did not apply until the evolution of sexually reproducing organisms some 2.5 to 3.5 billion years ago (Bernstein et al. 1981). The earth is 1 or 2 billion years older, the universe itself 10 billion years older. The conditions upon which causal structures depend are not equally well distributed in space and time. Biological causal structures are certainly more recent than some physical structures, and may be more ephemeral.

3.3. *Evolutionary Contingencies: Weak Contingency*

Even after the initial evolution of a structure and the associated causal dependencies that govern it, there may be changes in future environmental conditions that will break down those structures. With their demise the causal dependencies describing them will no longer apply. This is what Beatty dubs 'weak evolutionary contingency', i.e., that the conditions upon which the causal dependencies describing biological systems may change over time. They come and go. Thus there are types of historical contingency, or restrictions on the domain of applicability that attaches more often to biological regularities because life arose much later than other material forms, and may not hold on to its causal structures for as long. But universality never meant that the causal structure described in a law would occur at every point in space/time, rather that whenever and wherever the conditions for a relationship between the properties did occur, it would occur according to the relation described by the law.

3.4. *Strong Contingency (Multiple Outcomes Contingency)*

Beatty identifies a stronger sense of evolutionary contingency with "the fact that evolution can lead to different outcomes from the same starting point, even when the same selection pressures are operating" (Beatty 1995, p. 57). Here the focus is shifted from how likely or widespread are the conditions under which a causal relation will be operant to the uniqueness of the causal relationship or structure being evoked by those very conditions.

There are two ways in which strong contingency could occur. The first is when systems are indeterministic. If quantum processes have effects on macroscopic phenomena, then there can be cases where all the initial conditions are the same and the outcomes are nevertheless variant. Given mutations may be generated by radiation effects, this could well be symptomatic of biological systems. The second is in deterministic systems that are chaotic. Thus there may be the 'same' initial conditions, that is the same in so far as we can determine them to be the same, that nevertheless give rise to widely divergent outcomes. Complex biological systems are paradigm cases of chaos.

For deterministic cases, the *ceteris paribus* strategy for forcing this type of irregularity into the form of a law has been invoked (see Sober 1997). The reasoning is that if systems are deterministic there *must* be specific conditions, even if we can never know what they are or determine if they have occurred, that will distinguish the causal structures of the divergent outcomes. Sober argued that those conditions that caused the selection of a given structure and its accompanying regularities, e.g., the variations and fitness differences for the evolution of sexual reproduction and gametic segregation, could then form the antecedent of a strict law.

Beatty's own response to this type of maneuver is to suggest that there is no way to enumerate the conditions in the *ceteris paribus* antecedent clause to transform the biological generalizations that describe the causal structures of evolved biological systems. Earman, Roberts and Smith (this volume) remind us that this is only an epistemological worry. It may be that we will be able to articulate the conditions that completely specify the numerous conditions upon which some evolved structure depended and depends and hence be able to articulate the strict law governing that domain. If it is just a matter of knowing or not knowing all the conditions, then the very existence of *ceteris paribus* laws that cover evolutionary contingency is not called into question.

While it is clear how this strategy would work for weak contingency, how would it apply to strongly evolutionarily contingent regularities? Here, even when the causal conditions are operant, different, functionally equivalent outcomes are generated. If the relationship between the antecedent conditions and the evolved structure is probabilistic, then one could invoke the *ceteris paribus* strategy and allow for strict probabilistic laws. If the multiple outcomes are the result of a deterministic, but chaotic dynamics, then one could still claim that although we cannot discover the strict deterministic laws that direct the system into its separate outcome states, they nevertheless exist. Once again evolutionary contingent claims can be embedded via the specification of the contingent antecedent conditions in a strict law.

The problem I see with invoking the *ceteris paribus* response to Beatty's evolutionary contingency thesis is that it collapses the different ways in which a causal regularity can fail to be strict. By so doing it obscures, rather than illuminates the nature of biological knowledge.

4. COMPLEXITY AND CONTINGENCY

So far I have discussed the historical dimensions of chance and change characteristic of the evolved complexity of the domain of biological objects that make lawful behavior of the universal and exceptionless variety hard to come by. In addition there are contingencies that confound the strict lawfulness of currently existing complex biological systems. These are the result of the multi-level compositional structure of complex systems and the multi-component interactions at each level of those systems – complexity of both composition and of process.

4.1. *Multi-Level Interactions*

The operation of some causal mechanisms is contingent on the constraints imposed by their location within a multi-level system. For example, one could detail the causal relations describing the ovarian development in female bees, something that may well be a conserved trait from solitary ancestors to social descendents. The developmental laws that describe this process depend upon both the appropriate genes and internal (to the individual bee) conditions for the triggering of the expression of those genes at certain stages in the process of cellular specialization. However, when individuals come to live in social groups, the context in which these internal processes have to operate may change. When a female honeybee develops in a colony in which there is a queen, then the worker's ovarian development is suppressed by means of pheromonal control from the queen. If the queen should be killed, and the colony left without a queen, then the workers immediately begin to develop ovaries and produce haploid eggs. Thus the conditions upon which a causal mechanism operates depends upon the organizational structure in which it is embedded. This is not an incident peculiar to social groups.

The story of ovarian suppression in social insects is similar to the change of rules that occurs when the single celled stage of the slime mold ends with the aggregation to form the multi-cellular flowering stage I discussed above. Indeed, Buss (1987) has argued that the very origin of multi-cellular individuals from single-celled ancestors is one that involves the suppression of competition among the components (cells, or worker bees) of the new individual (multi-cellular individual, or colony) which permits the new individual to become the stable locus of evolutionary change. Thus the composition rules that characterize the various ways in which complex biological objects are formed will also affect the nature of the contingency in which the component mechanisms will operate.

4.2. *Multi-Component Causal Structures within a Level*

In addition to the type of contingencies arising with compositional hierarchies, there are also contingencies that affect specific causal mechanisms occurring within a level of organization. These are the result of the fact that most behaviors of complex biological systems are the result of the interaction of multiple mechanisms or causal factors. That there is more than one force acting to produce an outcome does not, by itself, threaten the existence, empirical accessibility, or usefulness of strict laws describing the individual components. Rather the problem arises with the nature of the interaction of these components. Positive and negative feedback loops, amplifying and damping interactions of a non-linear type are characteristic of complex biological systems.

For example, current research indicates that the behavior of individual foragers in a honeybee colony vary with genetic differences. This has been described in terms of the threshold levels for the stimulus required to initiate foraging behavior of individual bees. Experiments have supported the view that genetic variation accounts for behavioral variation via the genetic components determining individual threshold levels. However, it is now known that learning from the environment can also affect the behavior of foragers by moderating their threshold level. Indeed, the contribution of each of these very different mechanisms can amplify the expected probability of foraging behavior. Thus there may not be a 'regular' manner in which the contributions of different components generate a resultant outcome. Under some values of the components, it may be that the contribution of genetic variation is much stronger than variant learning experiences and hence completely determines the pattern of foragers in a colony or between colonies. Other times, the reverse might be true, and still other times, the interaction of the two components is operative in generating the pattern. Interactions may take different forms, including additive, swamping, damping and amplifying. Thus the operation of a single causal component, its contribution to the resultant effect, can be contingent not just on background standing conditions, but also on the other causal mechanisms operating at the same time. The nature of the contingency may vary with the different values that the variables in the component mechanisms take.

Redundancy and phase change phenomena can also be present in complex biological systems (see discussion above). These offer two more types of complex contingencies that have import for understanding the range of contexts in which the regular behavior of a set of variables may be disrupted. In systems with redundant processes, the contribution of any one may be elicited or moderated by the operation of another. For phase change, like non-linear dynamical processes, the nature of the function describing the behavior of variables may itself change under certain values of the variables or changes in external conditions.

5. THE PHILOSOPHICAL CONSEQUENCES

I have presented a variety of different ways in which causal dependencies in biology are contingent. This is damning of biology only if one retains the strict notion that laws must be universal and exceptionless. Instead, we can turn the question around and ask not whether biological claims can be transformed into strict laws, but rather when and how do biological claims perform the functions that laws are thought to serve? That is, how can less than strict laws explain, predict and assist in intervention? Recent work on causal dependence has done much to develop answers.

In his 2001 and his paper in this issue, Woodward applies his notion of explanation being grounded not by universal, exceptionless laws, but by generalizations that are invariant under intervention to cases in biology. He argues that his notion of invariance is both distinct from my idea of stability of conditions upon which relations hold and that invariance is what is needed for explanation, not stability. I want to explore Woodward's argument, to see where and why the disagreement occurs. First of all, there are many similarities in Woodward's and my account of laws. We both reject universality and exceptionlessness as necessary for knowledge claims to be deemed laws. We both want laws to have properties that come in degrees, rather than dichotomous values. What Woodward identifies as the relevant continuum is that of invariance (2000, p. 199) "...unlike lawfulness, invariance comes in gradations or degrees". For him, generalizations come with different degrees of invariance. It seems a bit odd to say invariance comes in degrees – since it seems to be the case that either the relationship between two variables is invariant or not, and how much it varies may be tracked, how much it fails to vary doesn't seem to make linguistic sense. However, what Woodward means does: he means that the relation between two variables is domain insensitive, such that if you change the value of the independent variable X, the dependent variable, Y, will stand in the same functional relationship say $Y = -kX + a$ for a range of changing values of X. There are however some values that X can take for which the function no longer is true. So the function is invariant under some but not all changes in the value of X. How many changes, how large the domain of invariance, of different functions will differ – and this is where I believe the degrees come in. Some functional relationships hold universally, some hold nearly universally – say except near a black hole – some hold for the majority of the time, and some hold some of the time. The value of X thus explains the value of Y by means of the functional relationship that describes the causal dependence of Y on X, in just those regions of the domain where the relationship is invariant. And for Woodward, having some domain of invariance is sufficient for explanation, even if it isn't universal, since some counterfactual situations – namely those changes in X where the function remains invariant – are supported.

I also attempt to provide a means of describing varying domains of applicability of different scientific laws (Mitchell 2000). To this end, I have identified a number of different continua that generalizations can be located within – in particular ontological ones of stability and of strength and representational ones of abstraction and cognitive accessibility. Ontological differences obtain between Fourier's law of thermal expansion, for example, and Mendel's law of segregation but it is not that one functions as a law and the other does not, or that one is necessary and the other is contingent. Rather one difference is in the stability of the conditions upon which the relations are contingent. Once the distribution of matter in the primordial atom was fixed, presumably shortly after the big bang, the function described by Fourier's law would hold. It would not have applied, if that distribution had been different, and indeed, will not apply should the universe enter a state of heat death. The conditions that both gave rise to the evolved structure of sexual reproduction and meiotic process of gamete production are less stable. The strength of deterministic, a probabilistic and a chaotic causal relations also vary.

There are methodological consequences to these variations in stability and strength. There is a difference in the kind of information required in order to *use* the different claims. It would be great if we could always detach the relation discovered from its evidential context, and be assured it will apply to all regions of space/time and in all contexts. But we cannot. Causal structures are contingent and, as I have argued above, they are contingent in a number of different ways. In order to apply less than ideally universal laws, one must carry the evidence from the discovery and confirmation contexts along to the new situations. As the conditions required become less stable, more information is required for application. Thus the difference between the laws of physics, the laws of biology, and the so-called accidental generalizations is better rendered as degrees of stability of conditions upon which the relations described depend, and the practical upshot is a corresponding difference in the way in which evidence for their acceptance must be treated in their further application.

Woodward compares his idea of invariance with my idea of stability and finds mine wanting for the job of explanation. Stability for me is a measure of the range of conditions that are required for the relationship described by the law to hold, which I take to include the domain of Woodward's invariance. However, stability can be a feature of relationships that are not invariant under ideal intervention. "Mere stability under some or even many changes is not sufficient for explanatoriness" (Woodward 2001). His counterexample is the case of common cause. According to Woodward, while the relationship between the two effects of a common cause is stable in any situation in which the common cause is operative, the one effect does not explain the other effect. Woodward's notion of invariance is supposed to capture this distinction, since some ideal interventions on the common cause system will show that a change in the value of the first effect will not be correlated with a change in the value of the second effect. The relationship breaks down, and so is not explanatory. If the world were such that those types of interventions never occurred naturally nor could be produced experimentally, on my view stability would be maintained, but invariance would still be transgressed, since there could be ideal situations in which it broke down. So, on his view, we would not have a 'law', on mine, we would.

The empiricist in me finds it difficult to detect the cash value of the difference Woodward is drawing between invariance and stability. If we could produce or witness the breaking of the relationship between the effects of a common cause, then we would find that the law describing the relationship between cause and either effect to be more useful than a law describing the relationship between effect and effect. Namely, the former would work in cases where the latter did not. But if there never were such cases to be found, then wouldn't they work equally well for prediction and intervention? If, on the other hand one requires a more substantial metaphysical warrant for explanation, as I believe Woodward does, then this constant conjunction or stable correlation would fail to explain why what happens, happens. The trouble is, I do not see how in practice you can distinguish the positions. If you have evidence for a common causal structure then on both accounts, the cause is a better predictor, and better at explaining the effects than either effect is of the other. If you have no evidence for a common causal structure, then the correlation, with the right sort of supporting evidence (like temporal order) would be taken to be explanatory.

I think the disagreement lies in the functions of laws on which we individually focus. Woodward admittedly attempts to account for only the explanatory function of laws. To perform this function, a generalization must report a counterfactual dependence. It has to describe a causal relationship that will remain true under certain episodes of "other things being different". It need not be true under *all* such episodes, i.e., exceptionless and universal, but to explain a particular occurrence, i.e., to say why a variable Y has the value it has by appeal to the value of a variable X, it must be the case that one could track x, y pairs in other circumstances, in particular, in interventions, *in the domain where the law was invariant*, and find the same relationship one finds in the explanandum situation. This is for Woodward what it means to say that Y is causally dependent on X and hence an occurrence or value of it is explained by appeal to X by means of the invariant generalization connecting X and Y. That is, Woodward lets domain restricted generalizations count as explanatory in just those domains where the relationship described in the generalization holds. Stability does just the same work, however it is weaker and includes what might turn out to be correlations due to a non-direct causal relationship. But for there to be a distinction between stability and invariance, then we would have to already know the causal structure producing the correlation.

Because the sciences I worry about embrace complexity, my goal has been to see how complexity affects the way we do science. Now if the world were hopelessly complex, or the dynamical evolution so rapid that we found ourselves in a Heraclitean Universe, we wouldn't be able to capture any knowledge that could be used downstream. But the history of science doesn't make this a plausible view – we can't be that deluded for that long that we actually can manipulate and predict events in the world. But it is equally misguided to take as an assumption that the world is simple, and expect to find that simplicity at every turn, and blame the investigator or impugn the science when simple laws are not to be found. We can learn about the features of a complex world, it is just not easy, and no single algorithm is likely to work in all the contexts in which complexity is found.

So where does that leave us with respect to the implications of complexity and contingency for our epistemological practices? First of all, descriptions of causal dependencies to be useful for prediction, explanation and intervention need not be universally true and exceptionless. As long as we can detail the domains in which the dependency is stable or invariant, then we can explain why what happens in that domain happens, and what will happen when changes in the magnitude of the causal parameters changes. However, there are different ways in which domains are restricted, or universality is lost including temporal and spatial restrictions that are the result of the evolutionary process; contextual restrictions in which certain parameter values or background conditions change the functions that describe the causal dependency; and contextual restrictions in which the operation of other causal mechanisms can interact in ways in which the effects of a cause are amplified, damped, made redundant or evoked. Representing all that variety of contingency by means of an unspecified *ceteris paribus* clause will mask the different strategies required to elicit information about complex contingencies in nature. In short, the context sensitivity of complex dynamical systems, like those studied by biology, entails a shift in our expectations. We should not be looking for single, simple causes. We should not be looking exclusively for universal causal relationships. And we must record and use not only the casual dependencies detected in a particular system or population to understand other systems and populations, but also the features that define the contexts present in the system under study. Without that information, exporting domain specific, exception ridden general truths cannot be done.

ACKNOWLEDGEMENTS

This research was supported by the National Science Foundation. Science and Technology Studies Program, Grant No. 0094395. It was improved by my participation in the "Working Group on Social Insects" organized by Robert E. Page, Joachim Erber and Jennifer Fewell and sponsored by the Santa Fe Institute and the serious discussion of earlier versions of the paper at the Max-Planck-Institut-für-Gesellshaftsforschung and at the

Greater Philadelphia Philosophy Consortium. I wish to thank Joel Smith, Jim Bogen and Clark Glymour for helpful comments.

REFERENCES

Beatty, J.: 1995, 'The Evolutionary Contingency Thesis', in G. Wolters and. J. G. Lennox (eds), *Concepts, Theories, and Rationality in the Biological Sciences*, University of Pittsburgh Press, Pittsburgh, pp. 45–81.

Beatty, J.: 1997, 'Why Do Biologists Argue Like They Do?', *Philosophy of Science* **64**, S432–S443.

Bernstein, H., G. S. Byers and R. E. Michod: 1981, 'The Evolution of Sexual Reproduction: The Importance of DNA Repair, Complementation, and Variation', *American Naturalist* **117**, 537–549.

Bonner, J. T.: 1988, *The Evolution of Complexity*, Princeton University Press, Princeton.

Brandon, R.: 1997, 'Does Biology have Laws: The Experimental Evidence', *Philosophy of Science* **64**, S444–S458.

Brandon, R.: 1982, 'Levels of Selection', in P. Asquith and T. Nickels (eds), *PSA 1982*, Vol. 1, Philosophy of Science Association, East Lansing, Michigan, pp. 315–323.

Brown, J. H. and G. West (eds): 2000, *Scaling in Biology*, Oxford University Press, Oxford.

Buss, L.: 1987, *The Evolution of Individuality*, Princeton University Press, Princeton, NJ.

Calderone, N. W. and R. E. Page, Jr.: 1992, 'Effects of Interactions among Genotypically Diverse Nestmates on Task Specializations by Foraging Honey Bees (*Apis mellifera*)', *Behavioral Ecology and Sociobiology* **30**, 219–226.

Cartwright, N. D.: 1994, *Natures Capacities and Their Measurement*, Oxford University Press, Oxford.

Cartwright, N. D.: 1999, *Dappled World: A Study of the Boundaries of Science*, Cambridge University Press, Cambridge.

Earman, J., J. Roberts, and S. Smith: 2002, '*Ceteris Paribus* Lost', (this issue).

Glymour, C.: 2001, *The Mind's Arrows: Bayes Nets and Graphical Causal Models in Psychology*, MIT Press, Cambridge, MA.

Goodwin, B. C.: 1994, *How the Leopard Changed its Spots: The Evolution of Complexity*, C. Scribner's Sons, New York.

Goodwin, B. C. and P. Saunders (eds): 1992, *Theoretical Biology: Epigenetic and Evolutionary Order from Complex Systems*, Johns Hopkins University Press, Baltimore, MD.

Gould, S. J.: 1990: *Wonderful Life: Burgess Shale and the Nature of History*, W. W. Norton, New York.

Horgan, J.: 1996, *The End of Science: Facing the Limits of Knowledge in the Twilight of the Scientific Age*, Broadway Books, New York.

Hull, D. L.: 1989, *The Metaphysics of Evolution*, State University of New York Press, Albany, NY.

Lange, M.: 2000, *Natural Laws in Scientific Practice*, Oxford University Press, Oxford.

Lange, M.: 2002, 'Who's Afraid of *Ceteris Paribus* Laws: Or: How I Learned to Stop Worrying and Love Them', (this issue).

Mitchell, S. D.: 1997, 'Pragmatic Laws', *Philosophy of Science* **64**, S468–S479.

Mitchell, S. D.: 2000, 'Dimensions of Scientific Law', *Philosophy of Science* **67**, 242–265.

Mitchell, S. D.: 2002, 'Contingent Generalizations: Lessons from Biology' in R. Mayntz (ed.), *Akteure, Mechanismen, Modelle. Zur Theoriefähigkeit makro-sozialer Analysen*, Max-Planck-Instituts für Gesellschaftsforschung.

Nicolis, G. and I. Prigogine: 1989, *Exploring Complexity: An Introduction*, W.H. Freeman, New York.

Page, R. E., Jr. and R. A. Metcalf: 1982, 'Multiple Mating, Sperm Utilization, and Social Evolution', *American Naturalist* **119**, 263–281.

Page, R. E., Jr. and S. D. Mitchell: 1991, 'Self Organization and Adaptation in Insect Societies', in A. Fine, M. Forbes, and L. Wessels (eds), *PSA 1990*, Vol. 2, Philosophy of Science Association, East Lansing, MI, pp. 289–298.

Pietroski and Rey: 1995, 'When Others Things Aren't Equal: Saving Ceteris Paris Laws from Vacuity', *British Journal for the Philosophy of Science* **46**, 81–110.

Raff, R.: 1996, *The Shape of Life*, University of Chicago Press, Chicago.

Salthe, S. N.: 1993, *Development and Evolution: Complexity and Change in Biology*, MIT Press, Cambridge, MA.

Simon, H.: 1981, *The Science of the Artificial*, 2nd edn, MIT Press, Cambridge, MA.

Sober, E.: 1984, *The Nature of Selection*, MIT Press, Cambridge, MA.

Sober, E.: 1997, 'Two Outbreaks of Lawlessness in Recent Philosophy of Biology', *Philosophy of Science* **64**, S432–S444.

Spirtes, P., C. Glymour, and R. Scheines: 1993, *Causation, Prediction, and Search*, Springer-Verlag, New York.

Tofts C. and N. R. Franks: 1992, 'Doing the Right Thing: Ants, Honeybees and Naked Mole-Rats', *Trends in Evolution and Ecology* **7**, 346–349.

Waters, C. K.: 1998, 'Causal Regularities in the Biological World of Contingent Distributions', *Biology and Philosophy* **13**, 5–36.

Wilson, E. O.: 1971, *Insect Societies*, Harvard University Press, Cambridge, MA.

Winston, M.: 1987, *The Biology of the Honey Bee*, Harvard University Press. Cambridge, MA.

Wimsatt, W.: 1986, 'Forms of Aggregativity', in Donagon, Perovich and Wedin (eds), *Human Nature and Natural Knowledge*, Reidel, pp. 259–291.

Woodward, J.: 2000, 'Explanation and Invariance in the Special Sciences', *The British Journal for the Philosophy of Science* **51**, 197–255.

Woodward, J.: 2001, 'Law and Explanation in Biology: Invariance is the Kind of Stability That Matters', *Philosophy of Science* **68**, 1–20.

Woodward, J.: 2002, *A Theory of Explanation: Causation, Invariance, and Intervention*, Oxford University Press, Oxford.

Zimmering, S., L. Sandler, and B. Nicoletti: 1970, 'Mechanisms of Meiotic Drive', *Annual Review of Genetics* **4**, 9–436.

Department HPS
1017 Cathedral of Learning
University of Pittsburgh
Pittsburgh, PA 15260
U.S.A.
E-mail: smitchel@pitt.edu

GERHARD SCHURZ

CETERIS PARIBUS LAWS: CLASSIFICATION AND DECONSTRUCTION

ABSTRACT. It has not been sufficiently considered in philosophical discussions of ceteris paribus (CP) laws that distinct kinds of CP-laws exist in science with rather different meanings. I distinguish between (1.) *comparative* CP-laws and (2.) *exclusive* CP-laws. There exist also *mixed* CP-laws, which contain a comparative and an exclusive CP-clause. Exclusive CP-laws may be either (2.1) *definite*, (2.2) *indefinite* or (2.3) *normic*. While CP-laws of kind (2.1) and (2.2) exhibit deductivistic behaviour, CP-laws of kind (2.3) require a probabilistic or non-monotonic reconstruction. CP-laws of kind (1) may be both deductivistic or probabilistic. All these kinds of CP-laws have empirical content by which they are testable, *except* CP-laws of kind (2.2) which are almost vacuous. Typically, CP-laws of kind (1) express invariant correlations, CP-laws of kind (2.1) express closed system laws of physical sciences, and CP-laws of kind (2.3) express normic laws of non-physical sciences based on evolution-theoretic stability properties.

1. INTRODUCTION: COMPARATIVE VERSUS EXCLUSIVE CETERIS PARIBUS LAWS

Philosophers of the last decades have repeatedly pointed out that most law statements, especially those in the non-physical sciences, do not express strict (i.e., universal and exceptionless) laws. Rather, they express so-called ceteris paribus laws, in short CP-laws.[1] The scientific 'dignity' of CP-laws, however, is a controversial matter.[2] In this paper I will try to show that "ceteris paribus" is a deeply *ambiguous* notion. It is better to differentiate the possible meanings before starting the attempt of explication. First of all, one should distinguish between two (families of) conceptions of CP-law: comparative versus exclusive.

The comparative sense of CP-clauses derives from the literal meaning of "ceteris paribus" as "the others being equal". A *comparative CP-law* makes an assertion about *functional* properties, henceforth called parameters.[3] It claims that the increase (or decrease) of one parameter, say f(x), leads to an increase (or decrease) of another parameter, say g(x), *provided* that all other (unknown) parameters describing the states of the underlying system(s) remain the same. Thus, a *comparative CP-clause* does not exclude the presence of other 'disturbing' factors, but merely

 Erkenntnis **57**: 351–372, 2002.
© 2002 *Kluwer Academic Publishers. Printed in the Netherlands.*

requires that they are kept constant. More precisely, a comparative CP-law compares the states of two systems of an underlying application class, one state where the parameter f has not been increased, and another state where the parameter f has been increased – and it requires both states to *agree* on all parameters which are causally *independent* from f (i.e., not affected by f). In particular, the quantitative parameters being compared may be the *probabilities* of some qualitative properties (expressed by predicates). Here are three examples:

(1)   Ceteris paribus, an increase of gas temperature leads to a (proportional) increase of gas volume (Gay-Lussac's gas law).

(2)   Ceteris paribus, increase of rain leads to an increase in growth of vegetation.

(3)   Ceteris paribus, (an increase of) alcoholization of the driver leads to an increased probability of a car accident.

While in (1) a *quantitative* relation between the increases is known (the relation of proportionality), in (2) only an *ordinal* relation between the increases is predicted (i.e., increase leads to increase). Finally, (3) is an example of a probabilistic comparative CP-law, where the consequent parameter g expresses a probability increase.

   In the philosophical debate, however, CP-laws have usually been understood in the different exclusive sense. An *exclusive CP-law* asserts that a certain state or event-type expressed by a (possibly complex) predicate *Ax* leads to another state or event-type *Cx provided* disturbing influences are *absent*. Ax is called the *antecedent* and Cx the *consequent* predicate. Thus, an exclusive CP-clause does not merely require keeping all other causally interfering factors constant; it rather *excludes* the presence of causally *interfering* factors. In agreement with this exclusive understanding, Cartwright has remarked that "the literal translation is 'other things being equal'; but it would be more apt to read 'ceteris paribus' as 'other things being right' " (1983, p. 45). Joseph (1980, p. 777) has spoken of "ceteris absentibus" clauses, and Hempel (1988, p. 29) calls exclusive CP-clauses "Provisos" ("...*provided* disturbing factors are absent"). Consider the following examples of exclusive CP-laws – (4) comes from physics and (5, 6) from psychology:

(4)   Ceteris paribus, planets have elliptical orbits (Lakatos op. cit.).

(5)   Ceteris paribus, people's actions are goal-oriented, in the sense that if person x wants A and believes B to be an optimal means for achieving A, then x will attempt to do B (Fodor, 1991; Dray 1957, pp. 132ff).

(6)   Ceteris paribus, frustration leads to aggression (Schurz 1995).

In (4), the CP-clause requires that other (non-negligible) forces on the planet except that of the sun are – not merely constant but – *absent*. Likewise, the CP-clause of (5) requires any factors causing irrational behavior to be absent. Note that (5) governs various special CP-laws, such as "CP people who want water try to get water" (Fodor 1991, p. 28). In (6), finally, the CP-clause excludes interfering factors of both psychological sort (e.g., depression) and physical sort (e.g., the influence of drugs).

The distinction between comparative and exclusive CP-laws is not disjoint. There are CP-laws which have both comparative and exclusive character, as in the following example from theoretical economy:

(7)   Ceteris paribus, an increase of demand leads to an increase of prices.

Not only must the compared economies agree in remainder factors; various interferes (such as political price regulations) must be excluded. We call these CP-laws *mixed* and treat them as (implicitly) governed by a comparative *and* an exclusive CP-clause; all what we say in the following about comparative and exclusive CP-clauses transfers to mixed CP-laws. The fact that comparatively formulated CP-laws are often mixed in character may explain why, historically, the two kinds of CP-laws have usually been conflated.

One may object to our distinction that some exclusive CP-laws can be reformulated in a comparative form, by interpreting *events* as *changes* in the values of certain parameters (cf. Gadenne 1984, p. 43f). In this way, the frustration-aggression law (6) may be reformulated as follows:

(6*)   Ceteris paribus, an increase of frustration leads to an increase of aggression.

But this reformulation does not at all diminish the difference. (6*) is *still* an exclusive CP-law, because interfering factors such as influences of certain drugs are not merely required to be constant, but must be absent. Otherwise, an increase of frustration will *not* lead to an increase of aggression. Generally speaking, if a CP-law "CP, if Ax then Cx" is truly

exclusive, then *even if* it is possible to reinterpret the predicates Ax and
Cx as changes of certain parameters Δf and Δg, the law will require in its
CP-clause not merely the non-occurrence of changes of other parameters;
it will also require that the values of other parameters are *close-to-zero* –
or more generally, that they are within an *uncritical* range where they are
not disturbing. This shows that the distinction between comparative and
exclusive CP-laws is *robust* as against different linguistic formulations.
As we shall see, this distinction will make a *decisive* difference for the
question of empirical content and testability.

2. COMPARATIVE CP-LAWS

In the following, we abbreviate a comparative CP-law as cCP(+ $\Delta f \rightarrow$
+Δg), where "cCP" stands for "ceteris paribus in the comparative sense".
(Likewise with decreases, where "+" is replaced by "−".) According to
the above analysis, the exact meaning of a cCP-law is defined as follows,
where "App" denotes the given application class of systems (x_1, x_2) being
compared, μ is a 2nd order variable ranging over (causally relevant) parameters, and "Ind(μ, λ)" is a 2nd order predicate standing for "μ is causally
independent of λ.".

(cCP-Def1): cCP(+ $\Delta f \rightarrow$ + Δg) *iff*

$\forall x_1, x_2 \in$ App: $(f(x_2) > f(x_1) \wedge \forall \mu(\text{Ind}(\mu, f) \rightarrow \mu(x_1) = \mu(x_2))) \rightarrow g(x_2) > g(x_1)$.

Thus, a cCP-law is (not a loose but) a *strict* implication. Its cCP-
clause is the second antecedent conjunct, $\forall \mu(\text{Ind}(\mu, f) \rightarrow \mu(x_1) = \mu(x_2))$,
which involves a universal 2nd order quantification; we abbreviate this
cCP-clause as cCP(x_1, x_2). Note that x_1 and x_2 may also be two different
temporal stages ($\langle x, t_1 \rangle$ and $\langle x, t_2 \rangle$) of the same system (x). For instance,
the precise meaning of (1) is that whenever temp(x_1) > temp(x_2) holds for
two gas probes which agree on all temp-independent parameters, vol(x_2)
> vol(x_1) will hold. Likewise for example (2).

The precise meaning of probabilistic cCP-laws such as (3) is expressed
in a slightly different form where one compares conditional probabilities.
Assume F and G express dichotomic properties and p is a suitably defined
statistical or 'objective' probability function. Then a probabilistic cCP-
law cCP(p(G/F) > p(G/¬F)) asserts that for all 'hypothetical' populations
P_1 and P_2 (in the sense of Eells 1991, p. 30ff) which agree on all F-
independent factors such that all P_1-members are F and all P_2-members are

not F, the probability of G in P_1 is greater than that of G in P_2. The agreement of P_1 and P_2 may be understood in two different senses, depending on whether one considers *pure* or (probabilistically) *mixed* circumstances. Assume, in the simplest case, $H_1, \ldots, H_m$ are all the remaining, relevant and F-independent, primitive dichotomic factors. Then $Q_1, \ldots, Q_n$ (where $n = 2^m$ and $Q_i = \pm H_1 \wedge \cdots \wedge \pm H_m$, "$\pm$" for "unnegated or negated") are all the possible pure circumstances. The possible mixed circumstances are defined as all possible probability distributions over the partition $Q_1, \ldots,$ Q_n (cf. Eells 1991, p. 85f). It follows from the law of mixed probabilities $(p(G/F) = \Sigma \{p(G/F \wedge Q_i).p(Q_i/F): 1 \leq i \leq n\})$ that a cCP law holds for all pure circumstances *iff* it holds for all mixed circumstances (Eells 1991, pp. 89ff). Summarizing, probabilistic cCP-laws can be defined as 2nd order generalizations as follows:

(cCP-Def2) $cCP(p(G/F) > p(G/\neg F))$ *iff for all* C, $p(G/F \wedge C) >$ $p(G/\neg F \wedge C)$ holds, where C ranges over circumstances in either the pure or the mixed sense.

In the terminology of *statistics*, probabilistic cCP-laws express circumstantially invariant fully partialized correlations. By the law of mixed probabilities we obtain the following equivalence:

(8) $cCP(p(G/F) > p(G/\neg F))$ *iff* for all hypothetical populations P_1, P_2: if (i) $p(F/P_1) > p(F/P_2)$ and (ii) P_1 and P_2 belong to the same cell C_i of the partition of circumstances, then $p(G/P_1) >$ $p(G/P_2)$.

This resembles definition cCP-Def1 and justifies the subsumption of non-probabilistic and probabilistic cCP-laws under the same category.

So far we have analyzed only *unrestricted*, or *pure* cCP-laws, which assert an invariant connection between Δf and Δg for all possible circumstances. Unrestricted *probabilistic* cCP-laws have been suggested by Cartwright (1989, p. 145f), Eells (1991, p. 85f) and others as an explication of generic *causal* relations. But unrestricted invariance claims are rarely true. Cartwright (1989, §5.2) mentions examples in classical physics. For example, if we compare two masses in situations which are exactly alike except that an additional gravitational force g is acting on mass 1, then mass 1 will experience an additional acceleration which is proportional to g/m. But this holds *only* because the classical total force law, f = m.a, equates the total force f with the *sum* of all component forces. In other words, the composition of forces is non-interactive. This point can be generalized: *unrestricted cCP-laws will only hold when the composition of causes is non-interactive* in the following sense:

(Def-Interact): Assume the dependent parameter $g(x)$ is a (total) function of *independent* parameters $f_1(x), \ldots f_n(x)$, i.e. for all x, $g(x) = h(f_1(x), \ldots, f_n(x))$; or in short: $g = h(f_1, \ldots f_n)$. Then $f_1, \ldots f_n$ are *non-interacting* (positive or negative) causes of g *iff* for all f_i and constant values a_j of the remaining parameters f_j ($1 \leq j \leq n$, $j \neq i$), the partial function $g(x) = h(f_i(x)/\forall j \neq i: f_j(x) = a_j)$ is (strictly) monotonic – either increasing, if f_i is a positive cause, or decreasing, if f_i is a negative cause.

Simple cases of non-interacting compositions are given if $h(f_1, \ldots, f_n)$ is the sum, or the product, of n one-placed functions $h_i(f_i)$ which are (strictly) monotonic. So, we do not speak of 'non-interacting' causes if some one-placed function $h_i(f_i)$ is not (strictly) monotonic, but, e.g., oscillating. In such a case we might say that the cause 'interacts with itself', since with increasing strength its character turns from a cause into a counter-cause. As an immediate consequence of (Def-Interact) we obtain:

THEOREM 1: Assume $g = h(f_1, \ldots f_n)$, as in (Def-Interact). Then: $cCP(+ \Delta f_i \rightarrow (+/-)_i \Delta g)$ holds for every f_i ($1 \leq i \leq n$) *iff* $f_1, \ldots f_n$ are non-interacting causes of g; where '$(+/-)_i$' = '+' ['−'] if f_i is a positive [negative] cause of g.

Proof: Because g is a (total) function of $f_1 \ldots, f_n$, all f_i-independent parameters which are causally relevant for g are exhausted by the parameters f_j, $j \neq i$. So (Def-Interact) and (cCP-Def1) imply that $f_1, \ldots, f_n$ are *non-interacting* causes of g exactly if $cCP(+ \Delta f_i \rightarrow (+/-)_i \Delta g)$ holds for every f_i.   Q.E.D.

As Cartwright herself has emphasized (e.g., 1983, p. 64ff) in systems of moderate complexity the composition of causes is usually *interactive*. In most situations we will therefore only have *restricted cCP-laws*, which assert the cCP-relation only for a restricted class of circumstances. In example (1), the cCP-relation between temperature and volume holds only for approximately ideal gases. In (2), it holds only under 'biologically normal' conditions. Also most probabilistic cCP laws, such as (3), will hold only for restricted classes of circumstances.

Restricted cCP-laws are mixed, exclusive-&-comparative CP-laws; so their prima facie form is $eCP(cCP(\pm\Delta f \rightarrow \pm\Delta g))$. Exclusive CP-clauses, in short *eCP-clauses*, will be investigated at length §§3–5; we will distinguish there between definite, indefinite and normic eCP-clauses. Accordingly, we can classify restricted cCP-laws into (i) definitely restricted cCP-laws (in which the restrictions of possible circumstances, or value-ranges, are clearly defined), (ii) indefinitely restricted cCP-laws, and (iii) normic cCP-laws (which assert that the cCP-connection normally holds).

Normic laws will be treated in §5. In §4 we will argue that indefinite eCP-laws, and hence also indefinitely restricted cCP-laws, are almost vacuous. In contrast, unrestricted and definitely restricted cCP-laws have empirical content and can be tested (provided their predicates denote observable properties). This will be shown in the remaining part of this section, by illustrating their role in *predictions* and *explanations*.

An unrestricted non-probabilistic cCP-law $cCP(+\Delta f \rightarrow +\Delta g)$ gives rise to the following deductive-nomological (D–N) argument pattern; it is stated in its prima facie form on the left and in its explicit form according to cCP-Def1 on the right; L is the law premise, A1,2 are the antecedent premises, and C is the conclusion:

$$L: cCP(+\Delta f \rightarrow +\Delta g) \qquad \forall x_1, x_2 \in App: (f(x_2) > f(x_1) \wedge \forall\mu(Ind(\mu, f)$$
$$\rightarrow \mu(x_1) = \mu(x_2))) \rightarrow g(x_2) > g(x_1)$$

$$A1: +\Delta f(a_1, a_2) \qquad f(a_1) > f(a_2)$$

$$A2: cCP(a_1, a_2) \qquad \forall\mu(Ind(\mu, f) \rightarrow \mu(a_1) = \mu(a_2))$$

$$\overline{C: +\Delta g(a_1, a_2) \qquad g(a_1) > g(a_2).}$$

If the conclusion C is known (or believed) beforehand, the argument pattern constitutes a D–N-*explanation* (of C), otherwise a D–N-*prediction* (of C). Note that the cCP-clause appears twice – uninstantiated as an antecedent conjunct of the cCP-law, and instantiated as the antecedent premise (A2). This is typical for *deductivistically* reconstructed CP-arguments, comparative as well as exclusive. Another typical feature is that the 'singular' CP-premise A2 is not genuinely singular but involves a 2nd order quantification over 'all other' (causally relevant) parameters. But – unlike the indefinite exclusive case – in the comparative case the singular cCP-clause $cCP(a_1, a_2)$ is experimentally realizable, *without the need to know* what these other relevant parameters are. Hence, the cCP-law is empirically testable.

A standard test method is the method of *randomized experiment* (Fisher 1951). In the present, non-probabilistic case one randomly chooses two application instances a_1, a_2 out of a homogeneous[4] *source* s, for example, two samples out of a chemical substance, and enforces an increased value of parameter f on one of them, say on a_1, by experimental manipulation – e.g., by heating one of the two substance samples. As a result, $f(a_1) > f(a_2)$ will hold and the two compared individuals a_1, a_2 will agree in all f-independent parameters, whatever they may be (modulo experimental errors, i.e., with high probability). So, the cCP-law predicts $g(a_1) > g(a_2)$, e.g., that the volume will increase (in the quantitative case it predicts $\Delta g(a_1, a_2)$ as a certain function of $\Delta f(a_1, a_2) \pm$ a random error term). If the predicted effect does not occur, this is a falsification or at least a strong disconfirmation of the cCP-law. The so-called 'holism' of falsification (i.e., the logical fact that if C is false we may give up either L or A1 or A2) is without practical significance, because by repeated random experiments we can guarantee A1 and A2 with arbitrarily high probability.

A negative result will strongly disconfirm the unrestricted cCP-law L as well as the source-restricted cCP-law L_s, which restricts the applications to the more narrow class App(s) $\subset$ App which contains only those instances x_i whose f- and μ-values lie in the value-range of the source. For example, if we performed our random experiment only for temperatures of 20–30 °C and pressures of 0.9–1.1 atm, then L_s assert the cCP-relation only for those pairs x_1, x_2 whose temperature- and pressure value lie in this region. The important point is that since L logically implies L_s (but not vice versa), a negative test result will strongly weaken L_s as well as L. This is not the case for *confirmation*. A positive test result will strongly confirm only the source-restricted law L_s, but not the unrestricted law L. It may well be that the asserted cCP-relation holds for the value-range of the source, but not for different value-ranges (e.g., it may fail for high temperatures, or low pressures). In order to confirm unrestricted CP-laws, series of experiments under maximally varying circumstances are needed. This presupposes theoretical knowledge about the remaining causally relevant factors. Since this knowledge is usually incomplete, confirmation claims regarding unrestricted cCP-laws have to be cautious.

The situation is essentially similar for probabilistic cCP-laws cCP(p(G/F) > p(G/¬F)), with the difference that the relation between premises and conclusion is now not deductive but probabilistically-inductive (in the sense of Hempel 1965, ch. 3.3–4). Out of a source population P, two random samples s_1, s_2 are taken, and factor F is imposed by manipulation on s_1 (the 'experimental group') but not on s_2 (the 'control' group). As a result, s_1 and s_2 differ in the frequency of F but agree in their frequency distribution over the F-independent factors, modulo random errors. In other words, s_1 and s_2 belong to the same cell in the partition of mixed circumstances, namely to that cell C_P to which also the source population P belongs. The cCP-law predicts that with high inductive probability, say ≥ 0.95, the difference in the G-frequency between s_1 and s_2 will be greater-or-equal to a certain significant threshold (see also Woodward 2002, §3). A negative test result will strongly disconfirm the unrestricted cCP-law $\forall C(p(G/F \wedge C) \geq p(G/\neg F \wedge C))$, as well as the P-restricted cCP-law $p(G/F \wedge C_P) \geq p(G/\neg F \wedge C_P)$, which is confined to the source population P. It is well-known that probabilistic random experiments can eliminate spurious correlations or spurious independencies but, again, they do not directly confirm unrestricted probabilistic cCP-laws (cf.

Eells 1991, p. 100f). In other words, a positive result will directly confirm only the P-restricted cCP-law $p(G/F \wedge C_P) \geq p(G/\neg F \wedge C_P)$; it might still be that in some different ('exceptional') circumstance C_k, $p(G/F \wedge C_k)$ $< p(G/\neg F \wedge C_k)$ holds. Thus, confirmation claims regarding unrestricted probabilistic cCP-laws have to be cautious, for the same reasons as in the non-probabilistic case.

3. EXCLUSIVE CP-LAWS: DEFINITE VERSUS INDEFINITE

In §2 we have seen that in order to obtain *correct* formulations of cCP-laws, one often needs an additional eCP-clause. A second important reason for the need of eCP-clauses is the search for predictively more efficient laws. For, the predictions made by cCP-laws (at least in their standard 'ordinal' form) are very weak: if classical physics were formulated with the help of cCP-laws, we could only make predictions about the *differences* between trajectories of (hypothetical) planets differing from each other in certain forces. In order to predict the trajectory of a planet, we need a list of *all* forces acting on it, and hence, we need an eCP-clause: "and nothing else".

   In what follows, we formulate an eCP-law *prima facie* as eCP(Ax → Cx), meaning that "if disturbing factors are absent, Ax leads to Cx". The variable (or n-tuple of variables) "x" is bound by the eCP-quantifier. As it was indicated in §2, the major distinction is that between definite and indefinite eCP-clauses and eCP-laws. A definite eCP-clause can be replaced by a (finite) list $D_1x, \ldots, D_nx$ *of possible disturbing factors* which are excluded in the antecedent of the eCP-law. Hence, a definite eCP-law can be deductivistically reconstructed as a strict implication of the form $\forall x((Ax \wedge \neg D_1x \wedge \ldots \wedge \neg D_nx) \rightarrow Cx)$. This transformation is called a *strict completion* of the eCP-law. It must be assumed thereby, of course, that the enriched antecedent predicate does not analytically imply the consequent predicate Cx.

   In most examples of eCP-laws such a strict completion is impossible. This is especially clear for our non-physical examples (5, 6) above. Here, the class of possible disturbing factors is completely heterogeneous and potentially infinite. The same is true for examples from biology (9) or technology (10):

(9)   eCP, birds can fly.

(10)   eCP, turning the ignition key starts the engine of my car.

It is a widely agreed that the real significance of eCP-laws lies in situations where a strict completion is impossible. This brings us to the critical part of this paper.

4. INDEFINITE eCP-LAWS: DEDUCTIVISTIC RECONSTRUCTIONS AND THEIR SHORTCOMINGS

In this section, I always understand "eCP" in the *indefinite* sense. Taken literally, an eCP-law makes a strict assertion *within the eCP-scope*: if disturbing factors are excluded, then Ax will *always* imply Cx (cf. also Pietroski and Rey 1995, p. 88). So it seems that an eCP-law can be reconstructed deductivistically as a strict implication of the form (i) "For all x: If 'eCP' and Ax, then Cx", or equivalently (ii) "For all x: If Ax, then Cx or else not 'eCP' ". Reconstruction (i), where an eCP-clause is understood as an *antecedent conjunct*, has been suggested by Lakatos' remarks in (1970, pp. 18, 26) – although Lakatos rejects this suggestion in his appendix (p. 98, fn. 3). It is also supported by Horgan/Tienson (1996, p. 138). Pietroski and Rey's definition (1995, p. 92, cond. ii) corresponds to the equivalent form (ii) (cf. Schurz 2001a). We express this reconstruction in definition (eCP-Def1a):

(eCP-Def1a): $eCP(Ax \to Cx)$ *iff* $\forall x((Ax \wedge eCP_{A \to C}(x)) \to Cx)$,

where Ax and Cx express nomological properties and $eCP_{A \to C}(x)$ stands for "situation x is eCP with respect to $Ax \to Cx$".

This definition gives us the opportunity to clarify the notion of an 'exception'. With a *strict* exception to an eCP-law we mean a true singular statement of the form E(a) which strictly falsifies it, and hence entails $Aa \wedge eCP_{A \to C}(a) \wedge \neg Ca$ (for some a). If we loosely speak of eCP-laws as laws which *admit exceptions*, then we do not mean strict exceptions but weak ones. We call a true singular statement E(a) a *weak* exception to an eCP-law iff E(a) entails $Aa \wedge \neg eCP_{A \to C}(a) \wedge \neg Ca$; in other words, E(a) implies that the eCP-clause is violated for case a (cf. Pietroski/Rey's 'abnormal' instances; 1995, p. 88).

The crucial question is how "$eCP_{A \to C}(x)$" can be defined. Several authors have tried to develop non-vacuous deductivistic reconstructions of eCP-laws. What their accounts have in common is that an indefinite eCP-clause is understood as a *second order quantification* which ranges over arbitrary nomological properties or events. A plausible characterization of "$eCP_{A \to C}(x)$" is to say that $eCP_{A \to C}$ is true in a situation x iff no nomological property or event-type ψ is present in x which causes $\neg Cx$, in the sense that it strictly and contingently implies it. With 'Nom(ψ)' for 'ψ is a nomological' and 'Cont(A)' for 'A is contingently true', we obtain the following definition (eCP-Def1b):

(eCP-Def1b): $eCP_{A \to C}(x)$ *iff* $\neg \exists \psi (Nom(\psi) \wedge \psi x \wedge Cont(\forall y(\psi y \to \neg Cy)))$.

Definition eCP-Def1a + b has two major defects which are informally explained as follows (cf. Schurz 2001a). *First*, a true eCP-law according to eCP-Def1a + b implies that, conditional on Ax, all ($\pm$Cx-events have deterministic causes, in the sense of strict contingent impliers (where "$\pm$" stands for "unnegated or negated"). For, if Aa holds, then either Ca is strictly determined by the absence of any disturbing condition, or $\neg$Ca is strictly determined by the presence of such a disturbing condition. *Second* and even worse, definition eCP-Def1a + b makes eCP-laws almost empty: an eCP-law does not only imply determinism w.r.t. (with respect to) the consequence predicate, it is also implied by it and, hence, is not stronger than it. For, whenever Aa is true (for arbitrary a), then either Ca and hence eCP(Aa $\to$ Ca) is true, or $\neg$Ca is true, in which case $\neg$Ca must have had a deterministic cause ψa, thus $eCP_{A \to C}(a)$ is false and so, eCP(Aa $\to$ Ca) is again true. The two defects are expressed in the following equivalence theorem which is proved in Schurz (2001a, Theorems 1 + 2).

THEOREM 2: Assume Ax and Cx are nomological predicates. Then: eCP(Ax $\to$ Cx) is true (according to eCP-Def1a + b) *iff* Cx has deterministic causes conditional on Ax.

Theorem (2.2) implies that because of their vacuity, eCP-'laws' may be completely *accidental*. To illustrate this counterintuitive consequence, assume that my and your actions are deterministically caused (conditional on a tautology). Then every eCP-law of the form "ceteris paribus, whenever I do X you do Y" is true, e.g., "CP if I play tennis you play chess", "CP if I play tennis, you don't play chess", etc.

Pietroski and Rey (1995) have suggested a refinement of definition eCP-Def1a + b. Their basic idea is the additional requirement that all exceptions to a CP-law must be explainable "by citing factors that are . . . independent" (p. 88), in the sense that besides explaining the exception they also explain some other evidence (p. 90). In Schurz (2001a; Theorems 3 + 4) it is proved that under the mild additional assumption that $\pm$Cx's deterministic causes are independently identifiable, Pietroski and Rey's definition also imply the counterintuitive consequences which are stated in Theorem 2.

The significance of Theorem 2 can be further demonstrated by ana-lyzing the role of indefinite eCP-laws in predictions and explanations. An eCP-law gives rise to the following D-N argument pattern, which appears in its prima facie form on the left, and in its explicit form according to eCP-Def1a + b on the right (L law premise, A1,2 antecedent premises, C conclusion):

L: $eCP(Ax \rightarrow Cx)$ $\forall x((Ax \wedge \neg\exists\psi(Nom(\psi) \wedge \psi x \wedge Cont(\forall y(\psi y \rightarrow \neg Cy)))) \rightarrow Cx)$

A1: Aa Aa

A2: $eCP(a)$ $\neg\exists\psi(Nom(\psi) \wedge \psi a \wedge Cont(\forall y(\psi y \rightarrow \neg Cy)))$

C: Ca Ca

As in the comparative case, the eCP-clause appears twice – uninstanti-ated as an antecedent conjunct of the cCP-law, and instantiated as the antecedent premise (A2), and as in the comparative case, the 'singular' CP-premise A2 is not genuinely singular but involves a 2nd order quanti-fication over 'all other' (causally relevant) parameters. But now – unlike the comparative case – there exists *no* procedure by which we can exper-imentally realize the eCP-clause eCP(a), i.e., by which we can guarantee the absence of 'disturbing factors whatever they are'. Since these disturb-ing factors are *not known* and may be of any sort, such a procedure is by definition impossible. Thus, there is no independent evidence for the truth of the singular eCP-premise *in advance*, before the outcome $\pm Ca$ has been observed. Observation of Ca will constitute forcing evidence for the truth of eCP(a), while observation of $\neg Ca$ will constitute forcing evidence for the falsity of eCP(a), as long as we believe in determinism w.r.t. $\pm Cx$. Indeed, this is exactly the content of Theorem 2, which tells us that the truth of the eCP-law is equivalent with the assumption of determinism.

This leads into a fatal trap of *immunization*: faced with an exception Aa $\wedge \neg Ca$ to an eCP-law we *must* infer that the exception was merely a weak one and, hence, we will *never* have reason to reject the eCP-law, as long as we believe in determinism w.r.t. $\pm Cx$. The only way out of this trap would be to formulate indefinite eCP-laws in a *weaker* way which does not presuppose determinism w.r.t. $\pm Cx$. But in this respect the defender of indefinite eCP-laws faces a *dilemma*. For the assumption of determinism w.r.t. $\pm Cx$ is one of the main *motivations* for formulating eCP-laws. If the lawlike connection between Ax and Cx asserted in eCP(Ax $\rightarrow$ Cx) is *weakened* to a statistical or normic connection so that even in the *absence* of any disturbing factor Aa will *not always* imply Ca, then the eCP-law is no longer adequate and should be replaced by a statistical or normic law. Indefinite eCP-laws face the dilemma that the very motivation for their formulation, namely the assumption of determinism, at the same time *undermines* them insofar it makes them almost vacuous.

These negative results are convergent with those of Earman et al. (2002) and Woodward (2002). I continue with a discussion of Fodor's account which, in the light of Woodward's sharp-written remarks, can be very brief. Fodor (1991, p. 28) starts from the assumption that the eCP-clause of psychological laws is *not* eliminable within the framework of psychology, but it could in principle be eliminated at the level of neurophysiology or physics. This eliminability of eCP-clauses is Fodor's defining criterion for eCP-laws as follows:

> **(eCP-Def2):**   eCP(Ax $\rightarrow$ Cx) is true *iff* every physical state φx which (minimally) realizes Ax has a strict completer ψx for Cx, where ψx is a strict completer of φx for Cx iff (i) $\forall$x(φx $\wedge$ ψx $\rightarrow$ Cx) is contingently true, but (ii) neither $\forall$x(φx $\rightarrow$ Cx) nor $\forall$x(ψx $\rightarrow$ Cx) are true (Fodor 1991, pp. 23).[5]

Fodor's definition eCP-Def2 is related to the previous definition eCP-Def1 by the following two points:

(1.) While according to eCP-Def1 an eCP-clause is a *universal* 2nd order quantification requiring the non-existence of disturbing factors, eCP-Def2 involves an *existential* 2nd order quantification requiring the existence of strict completers. However, in both definitions the quantifiers range over positive and negative properties. Hence, Fodor's completers may also express the absence of disturbers, and disturbers in the sense of eCP-Def1 may also include the absence of completers. Now assume that completers would have been defined simply by condition (i) of eCP-Def2; call them 'simple completers'. Then the assumption of determinism of $\pm$Cx conditional on Ax implies that given Ax, either a simple completer ψx (of Ax's realizer) for Cx is present, or a simple completer ψ^*x (of Ax's realizer) for $\neg$Cx is present, where a simple completer for $\neg$Cx is exactly a disturber in the sense of eCP-Def1. Hence, we obtain the following theorem (which we state without proof): eCP-Def1 and the simple-completer-version of eCP-Def2 are equivalent in the sense that *if* $\pm$*Cx has deterministic causes conditional on Ax, then a simple completer for Cx is present in x iff no disturber w.r.t. Cx is present in x.*

(2.) However, Fodor's definition of a completer ψx involves a relevance condition. It requires not only (i) that φx $\wedge$ ψx must be a sufficient condition of Cx, but also (ii) that φx as well as ψx must be necessary conjunctive parts of that condition. This relevance condition (ii) goes beyond eCP-Def1. As a result, Fodor's eCP-laws are slightly stronger than determinism w.r.t. $\pm$Cx conditional on Ax.

But this additional strength is still much *too weak* to prevent complete *accidentality* and almost-*vacuity*. For almost all arbitrary Ax and Cx one can find some (possible strange) circumstance ψx which together with Ax's realizer φx strictly and relevantly implies Cx, with the result that eCP(Ax → Cx) will be true according to eCP-Def2. As an illustration, assume Ax stands for 'person x looks to its right' and Cx for 'person x sees a kangaroo'. Let ψx be the circumstance 'a kangaroo stays at the right side of person x' (plus normal conditions for visual perception). Then every (minimal) Ax-realization φx has ψx as a possible completion for Cx such that both φx and ψx will be a necessary part of $\psi x \wedge \varphi x$. So "eCP(if a person looks to its right side, (s)he will see a kangaroo" is a true eCP-law according to Fodor's eCP-Def2. Another instructive example is given by Woodward (2002, §2).

The shortcomings of deductivistic reconstructions of indefinite eCP-laws can be summarized as follows:

(1.) Deductivistic reconstructions of indefinite eCP-laws are *too strong*, because they presuppose that the consequent predicate has deterministic causes. However, contemporary science makes it plausible that a certain portion of non-determinism does not only occur in quantum mechanics – it reigns in all domains of sufficiently complex systems (cf., e.g., Earman 1986).

(2.) At the same time, deductivistic reconstructions are *too weak*, because eCP-laws are almost empty and admit complete accidentality. In particular, eCP-laws imply nothing about the *statistical probability* with which undisturbed antecedent-cases will produce the consequent (cf. Pietroski/Rey 1995, p. 84; Schiffer 1991, p. 8). But this is counterintuitive. At least in the non-physical sciences, eCP-laws are usually asserted *only* if the situation without (non-neglectible) disturbing factors is also the *statistically normal* situation. Examples of deductivistically correct eCP-laws which are unintuitive because they violate this normality condition can be given in their thousands – e.g., CP no tire blows, CP there are no clouds in the sky, CP every human is naked (because that's how s/he was born), etc.

If the indefinite reconstruction of eCP-laws were the only alternative, then Schiffer (1991) and Earman et al. (2002) would be right that there simply *are* no genuine laws in non-physical sciences. Fortunately, we can offer an alternative. The above objections point to an alternative suggestion proposed by Schurz (1995) and, in different form, by Silverberg (1996): indefinite eCP-laws should be reconstructed as *normic* laws 'if Ax, then *normally* Cx', formalized as Ax ⇒ Cx, where ⇒ is a normic conditional operator which binds the variable x and obeys a *non-monotonic* (and hence non-deductivistic) logic. In this reconstruction, the indefinite eCP-clause is understood as a normality clause which does not make up a separate antecedent-conjunct or consequent-disjunct, but is *implicitly* contained in the normic conditional operator. In the next section I will try to show how the normic reconstruction avoids the two major problems of indefinite eCP-laws: lack of content and accidentality.

5. NORMIC LAWS IN NON-PHYSICAL SCIENCES

From now on we understand our non-physical examples (5,6,9,10) in the normic sense. Normic laws were discovered in the 1950's, when philosophers of history discussed Popper's and Hempel's model of deductive-nomological explanation. *Deductive* explanation requires strict laws. Dray pointed out that there are no strict laws in the historical sciences. Yet historians do explain. And when they do, they refer to normic rationality principles such as example (5) (Dray 1957, pp. 132–7). Scriven (1959) generalized this diagnosis to a broad range of sciences and introduced the name "normic laws". The dominant attitude at that time was to regard normic laws as being void of empirical content, because they are not (strictly) falsifiable (ibid.). This deductivistic attitude changed when philosophers examined more closely the nature of numerical-statistical laws. They too are not strictly falsifiable. And yet they *do* have empirical content, because they get *gradually disconfirmed* by the observation of significantly deviating sample frequencies. Just the same *gradual* (dis)confirmation strategy applies to normic laws, *provided* that they *imply* statistical normality claims. More precisely, provided that Ax $\Rightarrow$ Bx implies a high statistical probability of Bx conditional on Ax. I call this the *statistical consequence thesis*. The truth of this thesis would guarantee the empirical content and testability of normic laws.

However, severe objections have been raised against the statistical consequence thesis, by researchers in non-monotonic logic (e.g., McCarthy 1986) as well as by philosophers of science (e.g., Millikan 1984). According to these objections, a normic law asserts a certain kind of *prototypical* normality which is independent from statistical majority. The ability to fly, for example, is a property of *prototypical* birds, and this remains true even if, by some major disaster, the majority of birds were to lose their flying ability. I will now argue that prototypical normality is indeed *more* than statistical normality, and this "more" grants lawlikeness to normic generalizations; but nevertheless, prototypical normality implies statistical normality, in accordance with the statistical consequence thesis.[6] My argument is based on an *evolution-theoretic* foundation of normic laws which has been laid down in Schurz (2001b). For the present purpose I just need a brief sketch of it. All 'higher' sciences, from biology upwards, are concerned with living systems or with their cultural and technical products. What these systems have in common is the characteristic capacity of *self-regulation* under the permanent pressure of their environment. So this is my first thesis:

THESIS 1: Normic laws are the phenomenological laws of self-regulatory systems.

The identity of self-regulatory systems is governed by certain prototypical *norm states*, which these systems constantly try to approximate by means of their *real states*. They manage this with the help of *regulatory mechanisms* which compensate for *disturbing influences* of the environment. But what explains their proper or 'normal' functioning? The answer is contained in my next thesis:

THESIS 2: Almost all self-regulatory systems are *evolutionary systems* in the generalized 'Darwinian' sense. Their prototypical norm states and self-regulatory mechanisms have been gradually selected in their evolution history.

Evolution theory explains why evolutionary systems obey normic laws which imply high conditional statistical probabilities. The self-regulatory capacities of evolutionary systems are limited. Dysfunctions may occur, hence their normic behaviour may have various *exceptions*. Yet it must be the case that these systems are in their prototypical norm states in the high *statistical majority* of cases and times. For otherwise, they would not have *survived* in evolution. Put into a slogan: prototypical and statistical normality are connected by the law of evolutionary selection. Of course, this is a rather crude presentation; details can be found in Schurz (2001b). What is important in the present context: it is the explanation of normic laws in terms of an evolution history which (i) grants them lawlikeness, (ii) defines their excess content over mere statistical normality claims and (iii) at the same time explains how it comes that they imply statistical normality claims.

6. DIFFERENCES BETWEEN LAWS IN PHYSICS AND IN NON-PHYSICAL SCIENCES

Have *all* self-regulatory systems evolved by natural, cultural or technological evolution? Almost all – but there are exceptions. There also exist self-regulatory systems in nature which have come into existence by 'pre-evolutionary' processes, for example, the water-level of a lake, but their self-regulatory properties are so vulnerable that their normic regularities can hardly be called lawlike. So it is doubtful whether we can find normic *laws* in disciplines like physics or chemistry. But what is *then* the right analysis of eCP-laws in these areas? This is the question of the present section.

First of all, can we escape indefinite ceteris paribus clauses, or normic weakening, at least in *theoretical physics*, and achieve *genuinely strict* laws? That we can is the 'received' view. It has repeatedly been challenged by Cartwright and is strongly defended in Earman et al. (2002). In the next subsection I explain why I share Earman's opinion, though on slightly different reasons than given by him.

6.1. *Laws of Nature versus System Laws*

Laws of nature are those fundamental laws of physics which are not restricted to any special kinds of systems (be it by an explicit antecedent condition or an implicit application constraint). There are just a few of them. In classical physics, the *total force* law $f(x,t) = m(x).d^2s(x,t)/dt^2$ is a law of nature. It is a differential equation in which $f(x,t)$ figures as a *variable* function denoting the sum of *all* forces acting at time t on particle x without saying what these forces are. Another kind of laws of nature are *special force laws*, e.g. the classical laws for gravitational force or electric force – provided they are understood as laws about *abstract component forces* or 'capacities' in the sense of Cartwright (1989, pp. 183ff). Laws of nature are *strictly* true, *without* any ceteris paribus clause – but at the cost of being *per se* not *applicable* to *real* systems, because they do not specify *which* forces are active.

System laws, in contrast, refer to particular systems of a certain kind in a certain time interval Δt. They contain or rely on a *specification* of all forces which act within or upon the system x in the considered time interval Δt – the so-called *boundary conditions*. Examples of system laws in classical physics are Kepler's laws of elliptic planetary orbits, the law of free fall, etc. – *almost all* laws in physics textbooks are system laws. In physics, derivations of system laws from laws of nature and boundary conditions are usually not possible without simplification assumptions (cf. Cartwright 1983, pp. 104f). For non-physical system laws such as that birds can normally fly, derivation attempts of this sort are usually hopeless. System laws of this kind are usually obtained by purely empirical-inductive means. Because of their dependence on boundary conditions, system laws involve a certain portion of contingency. Nevertheless, they deserve the status of *lawlike* generalizations as well, because they support counterfactuals, such as "if you were to jump out of the window you would fall down", or "if this bird were to be hunted by a predator, it would fly away".

The importance of this distinction for our topic lies in the observation that while laws of nature go without eCP-clauses, system laws always need eCP-clauses. Laws of nature are silent about which forces are active. System laws require such a specification, and it is *here* where the eCP-clause enters the scene: *these* are the non-neglectible forces (or factors) and *nothing else*. This fact may shed light on some controversial matters. For example, philosophers like Schiffer (1991) or Earman et al. (2002) who think that non-physical sciences do not contain genuine laws of their own are right if one identifies "laws" with "laws of nature". But I think that this identification does not fit with the usage of the term "law" in science. What one rather should say is this: non-physical sciences do *not* have laws of nature of their own – but they *do* have system laws of their own.

Cartwright (2002, §3) argues that even the total force law has to be furnished by an eCP-clause. Her argument is connected with a consideration of Hempel (1988, p. 30). Hempel treats "total force" as a *theoretically indefinite* concept which refers to all logically possible kinds of forces, including even 'supernatural' (diabolic or telekinetical) forces. Earman and Roberts (1999, p. 444) are right in pointing out that Hempel himself does not conclude from this that his indefinite concept of total force *transcends* the resources of every physical theory. However, I think that this conclusion unavoidably follows. For, if Hempel's indefinite total force concept *were* a part of classical mechanics, then statements like "for all x: if a demonic force of quantity f but no other force acts on x at time t, then x experiences an acceleration proportional to f/m" would be consequences of classical mechanics. But this is clearly *not* the case. So, if "total force" were indeed an indefinite concept, then the only way to avoid this strange result would be Cartwright's suggestion (2002, §3), namely to furnish the total force law with an indefinite eCP-clause saying that "Provided no influences which are not describable as physical forces are active ...".

I think, the better way out of this dilemma is to understand total force as a *theoretically definite* concept, which is theoretically characterized by the total body of physical knowledge. Physical theories provide a theoretical classification of *all* kinds of elementary forces which is intended to be *complete*. At present, these are the four fundamental interaction forces (gravitational, electromagnetic, strong and weak nuclear force), and the inertial forces arising from collision. The concept of total force implicitly refers to this theoretical classification and, hence, excludes the *existence* of 'supernatural' forces of any sort. Therefore, the total force law is not in need of any indefinite eCP-clause.

6.2. *Physical System Laws as Theoretically Definite eCP-Laws*

System laws in physics such as our example (4) involve an eCP-clause: no non-negligible force acts on the planet except the gravitational force of the sun. What is the nature of this eCP-clause? It is certainly not a *normic* one, because planetary systems don't have self-regulatory capacities. Rather, the above analysis implies that this eCP-clause is a *theoretically definite* one, i.e., it is definite when expressed in the theoretical language of physics, where it refers to a supposedly complete theoretical classification of force kinds. More generally, my suggestion is to consider all idealized system laws of physics as theoretically definite eCP-laws. If this reconstruction is correct, then – in contrast to the non-physical sciences – the theoretical part of physics is governed by *strict laws*.

However, the *non-strict* part of physics has not been dissolved by these considerations. It reappears when we ask about the relation between the *idealized system* as described in the *theoretical antecedent description* (abbreviated as $A_t x$) and the *real system* as characterized by our empirical evidence – the *empirical antecedent description* (abbreviated as $A_e x$). As Hempel emphasizes in his remarks about "theoretical ascent" (1988, p. 21f), the inference from $A_e x$ to $A_t x$ must remain *uncertain*, for in the empirical or pre-theoretical language, a complete description of all possible disturbing factors is impossible. Do we face here again the problem of indefinite eCP-clauses? Or may we consider these application claims as holding with high conditional probability? According to Cartwright (1983, p. 47), the answer is *no*. For, the undisturbed ideal case described by $A_t x$ is a statistical rarity, and often even a physical impossibility (cf. Joseph 1980). However, there exist well-known methods of *approximation*, by which this challenge can be defeated. Without being able to enter this point in detail, I think that approximation procedures make it possible to turn an ideal theory into an empirical prediction which holds at least with high probability, conditional on $A_e x$.

6.3. *Closed (Isolated) versus Open (Self-regulatory) Systems*

The difference between eCP-laws in physical and non-physical sciences has its roots in the difference between closed or isolated systems of physics and chemistry, and open systems of non- physical sciences. Very generally, *systems* are physical ensembles composed of parts which preserve a relatively strict *identity in time*, by which they delimit themselves from their environment (Rapaport 1986, pp. 29ff). For closed systems this preservation of identity follows from their isolation which, in turn, is a matter of *postulate*: that our planetary system is stable is a *frozen accident* of cosmic evolution; should it be once devastated by a gigantic swarm of meteorites, then it stays so forever and will *not* regenerate. But how can we explain the relatively strict identity of *open* systems, which are permanently subject to possibly destructive influences from the environment? We have given the answer in §5: the stability of open 'living' systems follows from their self-regulatory capacities, which have arisen through evolution.

The difference between closed (or isolated) and open self- regulatory systems explains the different nature of their eCP-laws. For closed (or isolated) systems, a detailed specification of *all* forces ('and nothing else') is needed. For open self-regulatory systems, such a specification is neither possible nor necessary. It suffices to assume that the disturbing influences, whatever they may be, are within the manageable range of the system's self-regulatory compensation power. It is usually impossible to give an exact theoretical prediction of this 'manageable range' But evolution-theoretic considerations of §5 tell us that *normally* the external influences will be within this manageable range. This explains the *normic* character of the eCP-laws of open self-regulatory systems.

In conclusion, the differences between theoretically definite eCP-laws of physical sciences and normic eCP-laws of non-physical sciences over-whelm the similarities. Since, moreover, indefinite eCP-laws are almost empty pseudo-laws, it follows that the category of 'eCP-laws' is a very het-erogeneous one. More homogenous is the category of cCP-laws, which has been largely ignored in the philosophical discussion, with the exception of Woodward (2002, §3) who explicates ceteris paribus laws as probabilistic cCP-laws in my sense. But since most cCP-laws are restricted and hence partly exclusive in nature, similar problems and distinctions arise for them. In the upshot, it is recommendable to replace the category of ceteris paribus laws (simpliciter) by more differentiated notions.

NOTES

[1] E.g., Lakatos (1970, p. 17f), Cartwright (1983), Joseph (1980), Gadenne (1984), Hempel (1988), Fodor (1991), Pietroski/Rey (1995), Horgan/Tienson (1996, ch. 7), Kincaid (1996).
[2] E.g., Cartwright (ibid), Joseph (ibid), Scriven (1959), Schiffer (1991), Earman and Roberts (1999), Earman et al. (2002) and Woodward (2002).
[3] In what follows, we make use of terminological conventions of predicate logic: indi-vidual variables x, y ..., individual constants a, b ..., predicates F, G ..., function symbols f, g ...; Ax, Bx, ... denote formulas containing x (likewise for Aa, Ba, ...). We also allow x (a) to stand for n-tuples of individual variables (constants), and we allow for higher order variables and predicates.

[4] By "homogenous" I mean merely that the source consists of sufficiently many sufficiently small parts so that the process of taking samples can be regarded as a randomized experiment.

[5] Fodor has added to eCP-Def2 a second condition (1991, p. 27, cond. ii), but Mott (1992, p. 339f) has shown that this second condition does not work in the intended way; so we have omitted it in eCP-Def2. That Fodor's realizers have to be *minimal* ones was pointed out by Silverberg (1996) in reply to Mott (1992).

[6] This is substantiated by the fact that non-monotonic reasoning has an infinitesimal probability semantics (cf. Adams 1975, Spohn 2002) as well as a non-infinitesimal one (cf. Schurz 1998).

REFERENCES

Adams, E. W.: 1975, *The Logic of Conditionals*, Reidel, Dordrecht.

Cartwright, N.: 1983, *How the Laws of Physics Lie*, Clarendon Press, Oxford.

Cartwright, N.: 1989, *Nature's Capacities and their Measurement*, Clarendon Press, Oxford.

Cartwright, N.: 2002, 'In Favour of Laws That Are Not *Ceteris Paribus* After All', this issue.

Dray, W.: 1957, *Laws and Explanation in History*, Oxford University Press, Oxford.

Earman, J.: 1986, *A Primer on Determinism*, Reidel, Dordrecht.

Earman, J. and J. Roberts: 1999, '*Ceteris Paribus*, There Is No Problem of Provisos', *Synthese* **118**, 439–478.

Earman, J., J. Roberts and S. Smith: 2002, '*Ceteris Paribus* Lost', this issue.

Eells, E.: 1991, *Probabilistic Causality*, Cambridge University Press, Cambridge.

Fisher, R.: 1951, *The Design of Experiments*, Oliver and Boyd, Edinburgh.

Fodor, J.: 1991, 'You Can Fool Some of the People All of the Time', *Mind* **100**, 19–34.

Gadenne. V.: 1984, *Theorie und Erfahrung in der psychologischen Forschung*, Mohr, Tübingen.

Hempel, C. G.: 1965, *Aspects of Scientific Explanation and Other Essays*, Free Press, New York.

Hempel, C. G.: 1988, 'Provisos', in A. Grünbaum and W. Salmon, (eds), *The Limitations of Deductivism*, University of California Press, Berkeley, pp. 19–36.

Horgan, T. and J. Tienson:1996, *Connectionism and the Philosophy of Psychology*, MIT Press, Cambridge, MA.

Joseph, G.: 1980, 'The Many Sciences and the One World', *Journal of Philosophy* **77**(12), 773–790.

Kincaid, H.: 1996, *Philosophical Foundations of the Social Sciences*, Cambridge University Press.

Lakatos, I.: 1970, 'Falsification and the Methodology of Scientific Research Programmes', reprinted in I. Lakatos: 1978, *Philosophical Papers, Vol 1*, Cambridge University Press, Cambridge.

McCarthy, J.: 1986, 'Application of Circumscription to Formalizing Common-Sense Knowledge', *Artificial Intelligence* **13**, 89–116.

Millikan, R. G.: 1984, *Language, Thought, and Other Biological Categories*, MIT Press, Cambridge, MA.

Mott, P.: 1992, 'Fodor and Ceteris Paribus Laws', *Mind* **101**, 335–346.

Pietroski, P. and G. Rey:1995, 'When Other Things Aren't Equal: Saving Ceteris Paribus Laws from Vacuity', *British Journal for the Philosophy of Science* **46**, 81–110.

Rapaport, A.: 1986, *General System Theory*, Abacus Press, Cambridge, MA.

Schiffer, S.: 1991, 'Ceteris Paribus Laws', *Mind* **100**, 1–17.

Schurz, G.: 1995, 'Theories and their Applications – A Case of Nonmonotonic Reasoning', in W. Herfel et al. (eds), *Theories and Models in Scientific Processes*, Rodopi, Amsterdam, pp. 69–293.

Schurz, G.: 1998, 'Probabilistic Semantics for Delgrande's Conditional Logic and a Counterexample to his Default Logic', *Artificial Intelligence* **102**, 81–95.

Schurz, G.: 2001a, 'Pietroski and Rey on Ceteris Paribus Laws', *The British Journal for the Philosophy of Science* **52**,: 359–370.

Schurz, G.: 2001b, 'What Is 'Normal'? An Evolution-Theoretic Foundation of Normic Laws and Their Relation to Statistical Normality', *Philosophy of Science* **28**, 476–97.

Scriven, M.: 1959, 'Truisms as Grounds for Historical Explanations', in P. Gardiner (ed.), *Theories of History*, The Free Press, New York.

Silverberg, A.: 1996, 'Psychological Laws and Nonmonotonic Reasoning', *Erkenntnis* **44**, 199–224.

Spohn, W.: 2002, 'Laws, *Ceteris Paribus* Conditions, and the Dynamics of Belief', this issue.

Woodward, J.: 2002, 'There Is No such Thing as a Ceteris Paribus Law', this issue.

Chair of Theoretical Philosophy
University of Duesseldorf (Building 23.21)
D-40225 Duesseldorf, Germany
E-Mail: gerhard.schurz@phil-fak.uni-duesseldorf.de

WOLFGANG SPOHN

LAWS, *CETERIS PARIBUS* CONDITIONS, AND THE DYNAMICS OF BELIEF

ABSTRACT. The characteristic difference between laws and accidental generalizations lies in our epistemic or inductive attitude towards them. This idea has taken various forms and dominated the discussion about lawlikeness in the last decades. Likewise, the issue about *ceteris paribus* conditions is essentially about how we epistemically deal with exceptions. Hence, ranking theory with its resources of defeasible reasoning seems ideally suited to explicate these points in a formal way. This is what the paper attempts to do. Thus it will turn out that a law is simply the deterministic analogue of a sequence of independent, identically distributed random variables. This entails that de Finetti's representation theorems can be directly transformed into an account of confirmation of laws thus conceived.

1. PREPARATIONS

Laws are true lawlike sentences. But what is lawlikeness? Much effort went into investigating the issue, but the richer the concert of opinions became, the more apparent their deficiencies became, too, and with it the profound importance of the issue for epistemology and philosophy of science.

The most widely agreed prime features are that laws, in contrast to accidental generalizations, support counterfactuals, have explanatory power, and are projectible from, or confirmed by, their instances. These characteristics have long been recognized. However, the three topics they refer to - counterfactuals, explanation, and induction – were little elaborated in the beginning and are strongly contested nowadays. Moreover, the interrelations between these subjects were quite obscure. Hence, these features did not point to a clear view of lawlikeness, either. In this paper, I try to advance the issue. We shall see that the advance naturally extends to *ceteris paribus* laws, the general topic of this collection. Let me start with three straight decisions.

The first decision takes a stance on the priority of the prime features. I am convinced that it is the inductive behavior associated with laws which is the most basic one, and that it somehow entails the other prime features. I cannot justify this stance in a few lines. Suffice it to say that my study

 Erkenntnis **57**: 373–394, 2002.
© 2002 *Kluwer Academic Publishers. Printed in the Netherlands.*

of causation (1983) led me from Lewis' (1973) theory of counterfactuals over Gärdenfors' epistemic account of counterfactuals (cf., e.g., Gärdenfors 1981) ever deeper into the theory of induction where I finally thought I had reached firm ground. In Spohn (1991) I explained my view on the relation of induction to causation and thus to explanation. However, I did not return to counterfactuals (because I always felt that this subject is over-laid by many linguistic intricacies that are quite confusing). My decision finds strong support in Lange (2000) who starts investigating the relation between laws and counterfactuals and also arrives at induction as the most basic issue.

The second decision concerns the relation between laws and their prime features. When inquiring into lawlikeness the idea often was to search for something which *allows* us to use laws in induction, explanation and counterfactuals in the way we do. That is, given that induction is really the most basic aspect, lawlikeness should be something that *justifies* the role of laws in induction. This idea issued in perplexity; no good candidate could be found providing this justification.

There is an alternative idea, namely that lawlikeness is *nothing but* the role of laws in induction. In view of the history of inductive scepticism from Hume to Goodman – which made us despair of finding a deeper justification of induction and taught us rather to describe our inductive behavior and to inquire what is rational about it while being aware that this inquiry may produce only partial justification – this idea seems to be the wiser one. I do not mean to suggest that the lessons of inductive scepticism have been neglected; for instance, Lange (2000) endorses these lessons when explaining what he calls the root commitment concerning the inductive strategies associated with laws. But it is important to be fully aware of these lessons, and hence I shall pursue here the second idea and foreswear the search for deeper justifications. We shall see that we can still say quite a lot about rational induction.

We are thus to study the inductive properties of laws. This presup-poses some account of induction or confirmation within which to carry out the study. This is what my third decision is about. I think that on this matter philosophy of science went entirely wrong in the last 25 years. Bayesianism was always strong, and rightly so. In the 1950's and 60's much effort also went into the elaboration of a qualitative confirmation theory. However, this project was abandoned in the 70's. The main reason was certainly that the efforts were not successful at all. Niiniluoto (1972) gives an excellent survey that displays the incoherencies of the various attempts. An additional reason may be the rise and success of the theory of counterfactuals, which answered many problems in philosophy of science (though not problems of induction) and thus attracted a lot of the motivation originally directed to an account of induction.

In any case, the effect was that Bayesianism was more or less the only remaining well-elaborated alternative. This hindered progress, because deterministic laws and probability do not fit together well. Deterministic laws are not simply the limiting case of probabilistic laws, just as deterministic causation is not the limiting case of probabilistic causation. It is, for instance, widely agreed that the entire issue of *ceteris paribus* laws, to which we shall turn below, cannot find an adequate probabilistic explication. We find a parallel in the disparity between belief, or acceptance-as-true, and subjective probability, which was highlighted by the lottery paradox and has as yet not found a convincing reconciliation. My conclusion is, though I have hardly argued for it, that Bayesianism is of little help in advancing the issue of lawlikeness.

Philosophical logic was very active since around 1975 in producing alternatives, though not under the labels "induction" or "confirmation". However, these activities were hardly recognized in philosophy of science. Instead, they radiated to AI where they were rather successful. It is precisely in this area where we shall find help. Let me explain.

What should we expect an account of induction to achieve? I take the view (cf. Spohn 2000) that it is equivalent to a theory of belief revision or, more generally, to an account of the dynamics of doxastic states. This is why the topic is so inexhaustible. Everybody, from the neurophysiologist to the historian of ideas, can contribute to it, and one can deal with it from a descriptive as well as a normative perspective.

Philosophers, I assume, would like to come up with a very general normative account. Bayesianism provides such an account that is almost complete. There, rational doxastic states are described by probability measures, and their rational dynamics is described by various conditionalization rules. As mentioned, however, in order to connect up with deterministic laws, we should proceed with an account of doxastic states which represents plain belief or acceptance-as-true. Doxastic logic is sufficient for the statics, but it does not provide any dynamics. Probability < 1 cannot represent belief, because it does not license the inference from the beliefs in two conjuncts to the belief in their conjunction. Probability 1 cannot do it, either, because we would like to be able to update with respect to information previously disbelieved, because disbelieved propositions would have probability 0 according to this approach, and because Bayesian dynamics does not provide an account of conditionalization with respect to null propositions (that is why I called Bayesianism almost complete). Hence, Bayesianism is unhelpful. Belief revision theory (cf., e.g.,

Gärdenfors 1988) was deviced to fill the gap. Unfortunately, the dynamics it provides turned out to be incomplete as well (cf. Spohn 1988, sect. 3). There have been several attempts to plug the holes (cf., e.g., Nayak 1994 and Halpern 2001), but I still think that ranking theory, proposed in Spohn (1983, sect. 5.3, and 1988), though under a different name, offers the most convincing account for a full dynamics of plain belief.

In any case, this is my third decision: to carry out my study of the inductive behavior of laws strictly in terms of the theory of ranking functions. This framework may be unfamiliar, but the study will not be difficult, since ranking theory is a very simple theory. Still, there will be little space for broader discussion. Some of my results may appear trivial and some strange. On the whole, though, the study seems to me to be illuminating. But see and judge by yourself!

The plan of the paper is now almost obvious. In Section 2 I shall introduce the theory of ranking functions as far as needed. Section 3 explicates lawlikeness, i.e., the difference between laws and accidental generalizations insofar as it can be expressed in ranking terms. We shall see that this explication naturally leads to an inquiry of the role of *ceteris paribus* conditions and the like, a task taken up in Section 4. Since Section 3 analyzes belief in a law not as a belief in a regularity or some more sophisticated proposition, but rather as a certain inductive attitude, the immediate question arises how a law, i.e., such an inductive attitude, can be confirmed. This crucial question is addressed in Section 5. Section 6 will close with a few comparative remarks.

I thus focus entirely on the epistemological aspects of laws. I do not deny, but only neglect that laws have important metaphysical aspects as well. I have been less negligent in Spohn (1993), where I tried to understand causal laws as objectifications of inductive schemes, and in Spohn (1997), where I discussed both aspects of reduction sentences, the laws associated with disposition predicates. The two papers thus partially precede and partially transcend the present paper, and the unity of the three papers is less than perfect.

2. RANKING FUNCTIONS

Let us start with a set W of possible worlds, small rather than large worlds, as we shall see soon. Each subset of W is a truth condition or *proposition*. I assume propositions to be the objects of doxastic attitudes. Thus I take these attitudes to be intensional. We know well that this is problematic, and we scarcely know what to do about the problem. Hence, my assumption is just an act of front alignment.

The assumption also entails that we need not distinguish between propositions and sentences expressing them. Hence, I shall often use first-order sentences to represent or denote propositions and shall not distinguish between logically equivalent sentences, since they express the same proposition.

That is all we need to introduce our basic notion: κ is a *ranking function* (for W) iff κ is a function from W into **N** (the set of non-negative integers) such that $\kappa(w) = 0$ for some $w \in W$. For each proposition $A \subseteq W$ the *rank* $\kappa(A)$ of A is defined by $\kappa(A) = \min\{\kappa(w) \mid w \in A\}$ and $\kappa(\varnothing) = \infty$. For $A, B \subseteq W$ the *(conditional) rank* $\kappa(B \mid A)$ of B given A is defined by $\kappa(B \mid A) = \kappa(A \cap B) - \kappa(A)$. Since singletons of worlds are propositions as well, the point and the set function are interdefinable. The point function is simpler, but auxiliary; the set function is the one to be interpreted as a doxastic state.

Indeed, ranks are best interpreted as *grades of disbelief*. $\kappa(A) = 0$ says that A is not disbelieved at all. It does not say that A is believed; this is rather expressed by $\kappa(\bar{A}) > 0$,[1] i.e., that non-A is disbelieved (to some degree).[2] The clause that $\kappa(w) = 0$ for some $w \in W$ is thus a *consistency* requirement. It guarantees that at least some proposition, and in particular W itself, is not disbelieved. This entails the *law of negation*: for each $A \subseteq W$, either $\kappa(A) = 0$ or $\kappa(\bar{A}) = 0$ or both.

The set $C_\kappa = \{w \mid \kappa(w) = 0\}$ is called the *core* of κ (or of the doxastic state represented by κ). C_κ is the strongest proposition believed (to be true) in κ. Indeed, a proposition is believed in κ if and only if it is a superset of C_κ. Hence, the set of beliefs is *deductively closed* according to this representation.

There are two laws for the distribution of grades of disbelief. The *law of conjunction*: $\kappa(A \cap B) = \kappa(A) + \kappa(B \mid A)$, i.e., the grade of disbelief in A and the grade of disbelief in B given A add up to the grade of disbelief in A-and-B. And the *law of disjunction*: $\kappa(A \cup B) = \min\{\kappa(A), \kappa(B)\}$, i.e., the grade of disbelief in a disjunction is the minimum of the grades of the disjuncts. The latter is again only a consistency requirement, though a conditional one; if that law would not hold the inconsistency could arise that both $\kappa(A \mid A \cup B), \kappa(B \mid A \cup B) > 0$, i.e., that both A and B are disbelieved given A-or-B.

According to the above definition, the law of disjunction indeed extends to disjunctions of arbitrary cardinality. I find this reasonable, since an inconsistency is to be avoided in any case, be it finitely or infinitely generated. Note that this entails that each countable set of ranks has a minimum and thus that the range of a ranking function is well-ordered. Hence, the range **N** is a natural choice.[3]

However, here we would do better to avoid all complexities involved in infinity. Therefore I shall outright assume that we are dealing only with finitely many worlds and hence only with finitely many propositions. This entails that each world in W (or the set of its distinctive features) is finite in turn. Hence, as announced, they are small worlds. One may think that this is a strange start for an investigation of natural laws. However, an analysis of lawlikeness should work also under this finiteness assumption. After all, our world seems both to have laws and to be finite. Generalizing my observations below to the infinite case would require a separate paper.

There is no need here to develop ranking theory extensively. A general remark may be more helpful: ranking theory works in almost perfect parallel to probability theory. Take any probability theorem, replace probabilities by ranks, the sum of probabilities by the minimum of ranks, the product of probabilities by the sum of ranks, and the quotient of probabilities by the difference of ranks, and you are almost guaranteed to arrive at a ranking theorem. For instance, you thus get a ranking version of Bayes' theorem. Or you can develop the whole theory of Bayesian nets in ranking terms. And so on. The general reason is that one can roughly interpret ranks as the orders of magnitude of (infinitesimal) probabilities.

The parallel extends to the laws of doxastic change, i.e., to rules of conditionalization. Thus, it is at least plausible that ranking theory provides a complete dynamics of doxastic states (as may be shown in detail; cf. Spohn, 1988, sect. 5).

It is still annoying, perhaps, that belief is not characterized in a positive way. But there is remedy: β is the *belief function* associated with κ (and thus a belief function) iff β is the function assigning integers to propositions such that $\beta(A) = \kappa(\bar{A}) - \kappa(A)$ for each $A \subseteq W$. Similarly, $\beta(B \mid A) = \kappa(\bar{B} \mid A) - \kappa(B \mid A)$. Recall that at least one of the terms $\kappa(\bar{A})$ and $\kappa(A)$ must be 0. Hence, $\beta(A) > 0$, < 0, or $= 0$ iff, respectively, A is believed, disbelieved, or neither; and A is the more strongly believed, the larger $\beta(A)$. Thus, belief functions may appear to be more natural. But their formal behavior is more awkward. Therefore I shall use both notions.

Above, I claimed that a full dynamics of belief is tantamount to an account of induction and confirmation. So, what is confirmation with respect to ranking functions? The same as elsewhere, namely *positive relevance:* A *confirms* or is a *reason for* B relative to κ iff $\beta(B \mid A) > \beta(B \mid \bar{A})$, i.e., iff $\kappa(\bar{B} \mid A) > \kappa(\bar{B} \mid \bar{A})$ or $\kappa(B \mid A) < \kappa(B \mid \bar{A})$ or both.[4]

There is an issue here whether the condition should require $\beta(B \mid A) > \beta(B)$ or only $\beta(B \mid A) > \beta(B \mid \bar{A})$, as stated. In the corresponding probabilistic case, the two conditions are equivalent if all three terms are defined, but the first condition is a bit more general, since it may be defined while the second is not. That is why the first is often preferred. In the ranking case, however, all three terms are always defined, and the second condition may be satisfied while the first is not. In that case the second condition on which my definition is based seems to be more adequate.[5]

A final point that will prove relevant later on: Ranking functions can be mixed, just as probability measures can. For instance, if κ_1 and κ_2 are two ranking functions for W and if κ^* is defined by

$$\kappa^*(A) = \min\{\kappa_1(A), \kappa_2(A) + n\} \text{ for some } n \in N \text{ and all } A \subseteq W,$$

then κ^* is again a ranking function for W. Or more generally, if K is a set of ranking functions for W and ρ a ranking function for K, then κ^* defined by

$$\kappa^*(A) = \min\{\kappa(A) + \rho(\kappa) \mid \kappa \in K\} \text{ for all } A \subseteq W$$

is a ranking function for W. The function κ^* may be called the *mixture* of K by ρ.

This is all the material we shall need. I hope that the power and beauty of ranking theory is apparent already from this brief introduction. I have not argued here that if one wants to state a full dynamics of plain belief or acceptance-as-true, one must buy into ranking theory. I did so in Spohn (1988, sect. 3). Even that argument may not be entirely conclusive. However, I guess the space of choices is small, and I would be very surprised if a simpler choice than ranking theory were to be available.

Be this as it may, let us finally turn to our proper topic, the epistemology of laws.

3. LAWS

Let me start with a simple formal observation. Given some ranking function κ, to believe $A \wedge B$ means that $C_\kappa \subseteq A \cap B$, i.e., $\kappa(\neg A \vee \neg B) > 0$, i.e., $\min\{\kappa(\neg A), \kappa(\neg B)\} > 0$. This, however, can be implemented in many different ways. In particular, it leaves open how $\kappa(\neg A \vee \neg B)$ relates to $\kappa(\neg A)$ and $\kappa(\neg B)$ and thus whether or not $\kappa(\neg B \mid \neg A) = 0$. Hence, if you start with believing $A \wedge B$, but now learn that $\neg A$ obtains, you may, or may not, continue to believe B, depending on the value of $\kappa(\neg B \mid \neg A)$.

Basically the same point applies to believing a universal generalization. This, I propose, is the clue to understanding laws. Let us take $G = \bigwedge x(Px \to Qx)$ as our prototypical generalization ($\to$ always denotes material implication). I have already simplified things by assuming the worlds in W to be finite. This entails that the quantifier in G ranges over some finite domain D. For $a \in D$, let G_a be the instantiation of G by a, i.e., $G_a = Pa \to Qa$. Now to believe G in κ means that $C_\kappa \subseteq G$, i.e., $\kappa(\neg G) > 0$, i.e., $\min\{\kappa(\neg G_a) \mid a \in D\} > 0$. Thus, the generalization is believed as strongly as the weakest instantiation.[6]

Let us assume, moreover, that this is the only belief in κ, i.e., that $C_\kappa = G$; thus, no further beliefs interfere. This entails in particular that $\kappa(Pa \wedge Qa) = \kappa(\neg Pa \wedge Qa) = \kappa(\neg Pa \wedge \neg Qa) = 0 < \kappa(Pa \wedge \neg Qa)$ for each $a \in D$ and hence that $\kappa(\neg Qa \mid Pa) > 0$, i.e., that Pa is positively relevant for Qa. In other words, under this assumption the belief in the material implication $Pa \to Qa$ is equivalent to the positive relevance of Pa for Qa.

Again, the belief in G can be realized in many different ways. Let me focus for a while on two particular ways, which I call the "persistent" and the "shaky" attitude. If you learn about positive instances, G_a, G_b, etc. you do not change your beliefs according to κ, since you expected them to be positive, anyway.[7] The crucial difference emerges when we look how you respond to negative instances, $\neg G_a$, $\neg G_b$, etc. according to the various attitudes.

If you have the *persistent* attitude,[8] your belief in further instantiations is unaffected by negative instances, i.e., $\kappa(\neg G_b) = \kappa(\neg G_b \mid \neg G_a)$ ($b \neq a$), and indeed $\kappa(\neg G_b) = \kappa(\neg G_b \mid \neg G_{a_1} \wedge \cdots \wedge \neg G_{a_n})$ for any $n \in \mathbf{N}$ ($b \neq a_1, \ldots, a_n$). If, by contrast, you have the *shaky* attitude, your belief in further instantiations is destroyed by a negative instance, i.e., $\kappa(\neg G_b \mid \neg G_a) = 0$ and, a fortiori, $\kappa(\neg G_{\neq a} \mid \neg G_a) = 0$.[9]

The difference is, I find, characteristic of the distinction between lawlike and accidental generalizations. Let us look at two famous examples. First the coins:

(1)   All German coins are round.

(2)   All of the coins in my pocket today are made of silver.

It seems intuitively clear to me that we have the persistent attitude towards (1) and the shaky attitude towards (2). If we come across a cornered German coin, we wonder what might have happened to it, but our confidence that the next coin will be round again is not shattered. If, however, I find a copper coin in my pocket, my expectations concerning the further coins simply collapse; if (2) has proved wrong in one case, it may prove wrong in any case.

Or look at the metal cubes, which are often thought to be the toughest example:

(3)   All solid uranium cubes are smaller than one cubic mile.

(4)   All solid gold cubes are smaller than one cubic mile.

What I said about (1) and (2) applies here as well, I find. If we bump into a gold cube this large, we are surprised – and start thinking there might well be further ones. If we stumble upon an uranium cube of this size, we are surprised again. But we find our reasons for thinking that such a cube cannot exist unafflicted and will instead start investigating this extraordinary case (if it obtains for long enough).

As far as I see, this difference applies as well to the other examples prominent in the literature (cf., e.g., the overview in Lange 2000, pp. 11f.). However, my wording is certainly more determined than my thinking. According to my survey, intuitions are often undecided. In particular, the attitude seems to depend on how one came to believe in the regularity; there may be different settings for one and the same generalization. However, at the moment I am concerned with carving out what appears to me to be the basic difference. Therefore I am painting black and white. As we shall see, ranking theory will also allow for a more refined account.

In any case, what the examples suggest is this: We treat a universal generalization G as lawlike if we have the persistent attitude towards it, and we treat it as accidental if we have the shaky attitude towards it. Hence, the difference does not lie in the propositional content, it lies only in our inductive attitude towards the generalization or, rather, its instantiations.[10]

Given how much we have learned from Popper about philosophy of science, this conclusion is really ironic, since it says in a way that it is the mark of laws that they are *not* falsifiable by negative instances; it is only the accidental generalizations that are so falsifiable. Of course, the idea that the belief in laws is not given up so easily is familiar at least since Kuhn's days (and even Popper insisted from the outset that falsifications of laws proceed by counter-laws rather than simply by counter-instances). But I cannot recall having seen the point being stripped down to its induction-theoretic bones.

What I have said so far may provoke a confusion that I should hurry up to clarify. The persistent attitude towards $G = \bigwedge x(Px \rightarrow Qx)$ is characterized, I said, by the *independence* of the instantiations; experience of one instance does not affect belief about the others. In this way, belief about an instance G_b, i.e., the positive relevance of Pb for Qb, is persistent. But didn't we learn that one mark of lawlikeness is *enumerative induction*, i.e., the confirmation of the law by positive instances? Surely, enumerative induction outright contradicts the independence I claim.

Herein lies a subtle confusion. Belief in a law is more than belief in a proposition, it is a certain doxastic attitude, and that attitude as such is characterized by the independence in question. If I would have just this attitude, just this belief in a law, my κ would exhibit this independence. Enumerative induction, by contrast, is not about what the belief in a law *is*, but about how we may acquire or confirm this belief. The two inductive attitudes involved may be easily confused, but the confusion cannot be identified as long as one thinks belief in a law is just belief in a proposition.

However, what could it mean at all to confirm a law if it does not mean to confirm a proposition? Indeed, my definition in Section 2 applies only to the latter, and to talk of the confirmation of laws, i.e., of a second-order inductive attitude towards a first-order inductive attitude, is at best metaphorical so far; enumerative induction or falsificationism do not seem to make sense within this setting. In Section 5 I shall make a proposal for translating and saving enumerative induction and the falsification of laws. But here and in the next section I am concerned only with the attitude in which the belief in a law itself consists.

Is my explanation of lawlikeness a deep one? No, it is just as plain as, for instance, that of the counterfactual theorist who says that lawlikeness *is* support of counterfactuals or that a law *is* a universally quantified sub-junctive conditional. Analysis has to start somewhere, and it acquires depth only by showing how to explain other features of laws by the basic ones. That is a task that cannot be pursued here.[11] But I would like to insist that, as a starting point, the present analysis is to be preferred. There are good reasons for feeling uneasy about starting with subjunctives or a similarity relation between worlds. By contrast, ranking theory is a very plain theory with a very obvious interpretation.

The only doubt one may have about my starting point may concern its sufficiency as a basis of analysis. In particular one may feel that the crucial property of laws is one which justifies the inductive attitude I have described, say, some kind of material or causal necessity. Maybe. But I am sceptical and refer to my second decision in Section 1.

This does not mean that I have to sink into subjectivism, that I am bound to say that it is merely a matter of one's inductive taste what one takes to be a law. There may be objectivizations and rationalizations for our beliefs in laws. I do not intend to start speculating about this, but one very general rationalization is quite obvious. It is of vital importance to us to have persistent attitudes to a substantial extent. Something is almost always going wrong with our generalizations, and if we always had the shaky attitude, our inductions and expectations would break down dramatically and we could not go on living.

But of course, it is high time to admit that the distinction between the persistent and the shaky attitude is too coarse. It is not difficult, though, to gain a systematic overview within ranking theory. Let us see how many ways there are to believe the generalization G, i.e., for $\kappa(\neg G) > 0$. A natural and strongly simplifying assumption is

Symmetry: For all $a_1, \ldots, a_n, b_1, \ldots, b_n \in D$

$$\kappa(\neg G_{a_1} \wedge \cdots \wedge \neg G_{a_n}) = \kappa(\neg G_{b_1} \wedge \cdots \wedge \neg G_{b_n}).$$

In obvious analogy to inductive logic, symmetry says that the disbelief in violations of a generalization depends on their number, but not on the particular instances. For $n = 1$ symmetry entails that there is some $r > 0$ such that for all $a \in D$ $\kappa(\neg G_a) = \kappa(\neg G) = r$. More generally, symmetry entails, as is easy to see, that there is some function c from $\mathbf{N}$ to $\mathbf{N}$ such that for any $n + 1$ different $a_1, \ldots, a_n, b \in D$ the equality $\kappa(\neg G_b \mid \neg G_{a_1} \wedge \cdots \wedge \neg G_{a_n}) = c(n)$ holds, where $c(0) = r$. Indeed, all ranks of all Boolean combinations of the G_a are uniquely determined by the function c.

Another plausible assumption familiar from inductive logic is

Non-negative instantial relevance: For all $a_1, \ldots, a_n, a_{n+1}, b \in D$

$$\kappa(\neg G_b \mid \neg G_{a_1} \wedge \cdots \wedge \neg G_{a_n})$$
$$\geq \kappa(\neg G_b \mid \neg G_{a_1} \wedge \cdots \wedge \neg G_{a_n} \wedge \neg G_{a_{n+1}}).$$

This is tantamount to the function c being non-increasing.

Given the two assumptions there remain not so many ways to believe G; any non-increasing function c with $c(0) = r$ stands for one such way. Hence, the persistent attitude characterized by $c(n) = r$ for all n stands for one extreme, whereas the shaky attitude for which $c(n) = 0$ for $n \geq 1$ stands for the other. So, one may think about whether any ways in between fit the examples better than the extreme ones. Still, the consideration shows that the two attitudes I have discussed at length are suited best for marking the spectrum of possible attitudes.

4. OTHER THINGS BEING EQUAL, NORMAL, OR ABSENT

It is commonplace by now that laws or their applications are often to be qualified by some kind of *ceteris paribus* condition. As long as a law is conceived of as a proposition, the nature of this qualification is hard to understand. It seems to make the proposition indeterminate or trivial. But when we conceive of belief in a law as more than belief in a proposition, at least some mysteries dissolve in quite a natural way. Indeed, the account of laws given above almost yearns to be amended by such qualifications.

We should start, though, with the observation, often made in the literature, that we are dealing here with a mixed bag of qualifications. "*Ceteris paribus* condition" seems to have established itself as the general term, although it is clear to everyone that it really refers only to one kind of qualification. "*Ceteris paribus*" = "other things being equal" is obviously a relational condition. But what does it relate to? We shall return to this question. Another frequent qualification is that a law holds only in the absence of disturbing influences.[12] Still another way of hedging is to say that a law holds only under normal conditions.[13] A fourth kind are ideal conditions that are assumed by idealized laws though they are known not to obtain strictly. And there are other kinds, perhaps.

Yet another unclear thing is what exactly the qualifications are to act on. Some say it is the laws themselves that are hedged by the various conditions, while Earman and Roberts (1999) insist that the conditions exclusively pertain to the applications of laws to particular situations. Hence, provisoes in the sense of Hempel (1988, p. 151) which are "essential, but generally unstated, presuppositions of theoretical inferences" and hence part of the applications do not cover the phenomenon in full breadth, either.

This shows that the topic is not so uniform. Indeed, the inhomogeneity is a common theme in this collection. Still, let us squarely approach the topic from the vantage point reached so far. This will illuminate at least normal conditions and the absence of disturbing factors.

We have arrived at the result that the belief in the generalization $G = \bigwedge x(Px \rightarrow Qx)$ as a law is represented by having $\kappa(\neg G_a) > 0$ for each $a \in D$ in a persistent way, i.e., unshattered by violations of the law. I have praised persistence as a virtue. But, to be honest, does it not appear just narrow-minded? Violations of a law are cause for worry, not for stubbornness. Sure, but the worry should concern the violation, not the future. Indeed, ranking functions provide ample space for such worries. There may yet be a ramified substructure of additional conditions. Let me explain.

Suppose $\kappa(Pa) = 0$ and $\kappa(\neg Qa \mid Pa) = r > 0$, that is, you do not exclude Pa and believe Qa given Pa according to κ. This allows for there being an exceptional condition Ea such that $\kappa(\neg Qa \mid Pa \wedge Ea) = 0$. This is due to the non-monotonicity of defeasible reasoning embodied in a ranking function. Of course, this entails via the ranking laws that $\kappa(Ea \mid Pa) \geq r$, i.e., that the exceptional condition Ea is at least as strongly disbelieved as the violation of the law itself.

This, I find, is quite an appropriate schematic description of what actually goes on. We encounter a violation of a law, we are surprised, we inquire more closely how this was possible, and we find that some unexpected condition is realized under which we did not assume the law to hold, anyway. In this way, hence, each ranking function representing the belief in the law G automatically carries an aura of *normal conditions* which is implicit at the level of belief, i.e., the function's core, and becomes explicit only if we look more deeply at the substructure below the core. This substructure may indeed dispose to further changes of opinion. There may, e.g., be a further condition $E'a$ such that the law G is reinstalled, i.e., $\kappa(\neg Qa \mid Pa \wedge Ea \wedge E'a) > 0$ for all $a \in D$. Defeasible reasoning may have arbitrarily many layers according to a ranking function.

Relative to a given κ embodying the belief in the law G we can even define the normal conditions hedging G. For, if Ea and Fa are exceptional conditions, $Ea \vee Fa$ is so as well. $\kappa(\neg Qa \mid Pa \wedge Ea) = \kappa(\neg Qa \mid Pa \wedge Fa) = 0$ is easily seen to imply $\kappa(\neg Qa \mid Pa \wedge (Ea \vee Fa)) = 0$. Hence, the disjunction E^* of all exceptional properties E for which $\kappa(\neg Qa \mid Pa \wedge Ea) = 0$ for all $a \in D$ (or for some $a \in D$, if symmetry is given) is the *weakest* exceptional property, and we may thus define the normal conditions N^* pertaining to G (relative to κ) as the complement or negation of E^*.

Note that N^* is not simply the disjunction N of all maximal properties M such that the law G holds given M, i.e., $\kappa(\neg Qa \mid Pa \wedge Ma) > 0$ for all $a \in D$. N^* is at least as strict as N and usually stricter. For instance, the condition $E \wedge E'$ under which the law G was assumed to be reinstalled two paragraphs above would be a specification of N, but not of N^*. The example also shows that normal conditions are more adequately explicated by N^*, because the condition $E \wedge E'$ should count as doubly exceptional and indeed counts as exceptional according to N^*, whereas it would count as normal according to N.

In any case, I find it entirely appropriate that normal conditions are thus explicated relative to a given doxastic state. Normalcy is something in the eye of the observer, in the first place, and therefore it is best described via its epistemic functioning. And ranking functions are particularly suited to grasp this.

However, this specifies only the statics of normal conditions. But we are rather interested in their dynamics, i.e., in the way in which our conception of them changes. After all, if we encounter a violation of a law, closer inspection of the case will often not confirm our previous understanding of exceptions, but will instead inform and revise it. This issue, however, belongs under the heading "confirmation of laws", which I address only in the next section.

So much for the ramifications of the belief in a single law G. The next issue to face, hence, is: How to believe in several laws at once, in particular if they pertain to the same property? Let us look at the simplest example: Often we seem to believe in the law $G = \bigwedge x(Px \rightarrow Qx)$ *and* in a further law $G' = \bigwedge x(P'x \rightarrow \neg Qx)$ predicting *non-Q* for circumstances P'.[14] How can we do this?

This is the problem of the *superposition of laws* or, if the laws are causal, of the *interaction of causes*.[15] In mechanics the problem finds an elegant solution: the total force acting on a body is just the vector sum of the individual forces, each of which is governed by a specific force law. But in general there is no general solution. Only so much can be said:

It is possible to believe both in G and G', though only if one also believes that $\neg \bigvee x(Px \wedge P'x)$. This is simply a matter of logic.

From the ranking perspective two remarks must be added. First, both laws can also be believed in the sense explained here, but only if the disbelief in each instance $Pa \wedge P'a$ is sufficiently strong. Second, and more importantly, even if a ranking function κ represents the belief in both G and G' as laws it still contains a prediction for the unexpected case that a instantiates both P and P'; $\beta(Qa \mid Pa \wedge P'a)$ must take some value. Hence, if two competing laws are believed in κ, they are automatically superposed in κ in some way (which may well be suspension of judgment, i.e., $\beta(Qa \mid Pa \wedge P'a) = 0$).

Even though this description is very unspecific (and is bound to be so), there is one point where it seems to be false. The description assumes that for each law it is exceptional in the above sense that the other law applies as well in a given case. But this is not how we normally look at the laws. We should be able to account for the superposition of G and G' even if $\kappa(Pa \wedge P'a) = 0$. This is why the present problem cannot be subsumed under the problem of normal conditions. But what else could be the account?

The only way seems to be to make the laws exclusive, i.e., to modify G into $\bigwedge x(Px \wedge \neg P'x \rightarrow Qx)$ and G' into $\bigwedge x(P'x \wedge \neg Px \rightarrow \neg Qx)$ and to modify κ correspondingly. The laws did not make any prediction for the case $\neg Pa \wedge \neg P'a$, anyway. What is left open, hence, is the case $Pa \wedge P'a$, for which one may, and has to, assume some degree of (dis-)belief in Qa. The resulting κ, according to which three laws, the modified G and G' and the new one, are believed, may also be called a superposition of the laws G and G'.[16] This consideration shows that the belief in a law as such, as I have described it, is implicitly understood in abstraction from other things, i.e., other relevant laws, and this abstraction is made explicit in the superposition in the second sense; i.e., in the modifications of G and G'.[17]

So, in which way do these remarks bear on the hedgings of laws familiar from the literature? Let me briefly summarize.

The account of normal conditions I have given is exactly the one compellingly suggested by the literature on non-monotonic reasoning, default logic, or whatever the labels were, which has been richly produced since 1975. What I add is only the conviction that ranking theory, owing to its completeness concerning induction or belief revision, provides the optimal base for studying these phenomena.

The absence of disturbing influences or factors may stand for various things. It may simply mean the presence of normal conditions. Or it may mean that the case at hand is not governed by a further law which would require some guess or knowledge as to how the laws involved superimpose. To this extent, at least, this kind of hedge is covered by my remarks.

What about *ceteris paribus* clauses? As already mentioned, they require a standard of comparison which is usually left implicit. The default standard, I guess, is given by the normal conditions. In this case, other things being equal just means other things being normal. If, however, the standard of comparison is taken as variable, then the clause yields what Schurz (2002) calls comparative CP-laws, or it amounts to some such principle like "equal causes, equal effects" or "induction goes by suchnesses, not thisnesses" which might be explicated by symmetry principles like the one above. But I shall not pursue this issue.

Finally, I have not said anything about idealizations. This seems to be a somewhat different topic. But I should at least mention that it is accessible to the belief revision perspective as well, as has been shown by Rott (1991).

5. ON THE CONFIRMATION OF LAWS

At several crucial points we missed an account of the confirmation of laws, and it was quite unclear how to give one, since the issue is not about the confirmation of propositions, which was already well handled by ranking functions. My paper would be badly incomplete without such an account.

But I have a proposal. Indeed, it will not be a surprise to anyone who is aware of the close similarity between probability and ranking theory, who has in particular noticed that a law according to my conception is analogous to a sequence of independent, identically distributed random variables, and who knows the work of de Finetti (1937). In his famous theorems de Finetti showed that there is a one-one-correspondence between symmetric probability measures for an infinite sequence of random variables and mixtures of Bernoulli measures according to which the variables are independent and identically distributed, and that the mixture concentrates more and more on a single Bernoulli measure as evidence accumulates. He thus showed to the objectivist that subjective symmetric measures provide everything he wants, i.e., beliefs about statistical hypotheses that converge toward the true one with increasing evidence.

The issue between objectivism and subjectivism is not my concern. Ranking functions are thoroughly epistemological and have as such no objective interpretation.[18] Still, we can immediately extract an account of the confirmation of laws from de Finetti's theory. Since this will look a bit artificial and formalistic, I shall demonstrate this with the basic construction and not discuss variants and ramifications.

Let us start with n mutually exclusive and jointly exhaustive properties or predicates $Q_1, \ldots, Q_n$ (these are Carnap's Q-predicates). For each $i \leq n$ we have the elementary law $G^i = \neg \bigvee x Q_i x = \bigwedge x \neg Q_i x$. For any proposition $A \subseteq W$ we may now count how often the law G^i is violated if A obtains; this is done by the function $v(A, i) = \text{card}\{a \in D \mid A \subseteq Q_i a\}$.[19] So, if we define the ranking function κ^i for W by $\kappa^i(A) = v(A, i)$, κ^i precisely represents the belief in the law G^i. Without any evidence, though, we do not believe in any law G^i. Our attitude towards the laws is rather represented by the ranking function ρ_0 for which $\rho_0(\kappa^i) = 0$ for each $i = 1, \ldots, n$. Hence, our doxastic attitude towards the propositions $A \subseteq W$ is represented by the mixture κ_0 of the κ^i with respect to ρ_0, as defined by

$$\kappa_0(A) = \min_{i \leq n} \kappa^i(A) + \rho_0(\kappa^i) = \min_{i \leq n} v(A, i).$$

Now, how does this attitude change by experience? Via conditionalization, as always. But let us describe this in detail. Let $\mathbf{r} = \langle r_1, \ldots, r_n \rangle$ stand for any sequence of n non-negative integers, and let $r = r_1 + \cdots + r_n$. Define next $E(\mathbf{r})$ to be the proposition (evidence) that among the first r objects precisely r_i instantiate Q_i ($i = 1, \ldots, n$); the order of instantiation is irrelevant. Clearly, $\kappa_0(E(\mathbf{r})) = \min r_i$. Let B range over propositions about the remaining objects and not the first r ones, and let $\kappa_{\mathbf{r}}$ be the ranking function that we have for those propositions after receiving evidence $E(\mathbf{r})$. Then we have:

$$\kappa_{\mathbf{r}}(B) = \kappa_0(B \mid E(\mathbf{r})) = \kappa_0(B \cap E(\mathbf{r})) - \kappa_0(E(\mathbf{r}))$$
$$= \min_{i \leq n}(v(B, i) + r_i) - \min_{i \leq n} r_i = \min_{i \leq n}(\kappa^i(B) + (r_i - \min_{i \leq n} r_i)).$$

That is, if we define $\rho_{\mathbf{r}}$ by $\rho_{\mathbf{r}}(\kappa^i) = r_i - \min_{i \leq n} r_i$, we have

$$\kappa_{\mathbf{r}}(B) = \min_{i \leq n}(\kappa^i(B) + \rho_{\mathbf{r}}(\kappa^i)).$$

Hence, $\kappa_\mathbf{r}$ is the mixture of the κ^i with respect to $\rho_\mathbf{r}$. So, the evidence $E(\mathbf{r})$ makes us change our attitude towards the laws from ρ_0 to $\rho_\mathbf{r}$, and $\rho_\mathbf{r}$ represents the degrees to which the various laws have been confirmed or rather disconfirmed. If $\rho_\mathbf{r}(\kappa^i) > 0$, we might say that κ^i is falsified, but note that falsification is never conclusive in this construction.

This account is essentially a translation of de Finetti's results into the framework of ranking functions. I find the translation basically plausible, and it strongly suggests following its course. One should characterize the class of ranking functions which represent mixtures of laws, and one should inquire the extent to which the representation is unique (for instance, there is an obvious one-one-correspondence between the $\kappa_\mathbf{r}$ and the $\rho_\mathbf{r}$ in the above mixtures). One should look at de Finetti's representation results for the infinite as well as for the finite case (recall the finiteness assumption made in this paper). The ranking analogue to de Finetti's notion of partial exchangeability would be particularly interesting. And so forth.[20]

On the other hand, the translation still looks artificial and quite detached from actual practice. For instance, if min r_i is large, one would tend to say that all of the laws G^i are disconfirmed by $E(\mathbf{r})$ and to conclude that none of the laws holds. One might account for this point by defining some κ^0 representing the belief in lawlessness, by mixing it into κ_0, say with the weight $\rho_0(\kappa^0) = s$, and by finding then that as soon as min $r_i > s$ we have $\rho_\mathbf{r}(\kappa^i) = 0$ only for $i = 0$. Moreover, one might wonder how precisely this story of mixtures carries over to the belief in a given law and its possible hedgings by various possible normal conditions, since one would like to be able to account for one hedging rather than another being confirmed by the evidence. And so on.

All this shows that there is a lot of work to do in order to extend the proposal and to apply it to more realistic cases. Still, the message should be clear already from the case I have explained in detail. The theory of mixtures provides a clear account of what it means to confirm and disconfirm not only propositions, but also inductive attitudes such as ranking functions representing belief in laws. Hence I was not speaking metaphorically when I talked about such confirmation earlier in the paper.

6. SOME COMPARATIVE REMARKS

The literature on *ceteris paribus* laws is rich and disharmonious, and so far I have only added to the polyphony. Since the idea of this ERKENNTNIS issue was to promote harmony (which does not require everybody to play the same melody), I should close with some comparative remarks.

So far, Schurz (1995) and Silverberg (1996) were the only ones to decidedly use the resources of non-monotonic reasoning for our topic (cf. also Schurz 2002, sect. 5). I emphatically continue on this line of thought, but we certainly have an argument about the most suitable account of non-monotonic reasoning.

What is novel to me is that the topic may also be approached from the learning-theoretic perspective. Indeed, I feel that Glymour (2002) and the present paper sandwich, as it were, the paper by Earman, Roberts, and Smith (2002), which is the focal challenge of this collection. How the two sides stick together is not clear. However, Kelly (1999) has established a general connection between formal learning theory and ranking theory, and the relation should become closer when one compares Kelly (2002) with the present Section 5. So, let me briefly sketch my part of the pincer movement towards Earman et al. (2002), which will lead me across some other positions.

Clearly, my position is very close to that of Lange (2000), who says, for instance, that "the root commitment that we undertake when believing in a law involves the belief that a given inference rule possesses certain objective properties, such as reliability" (p. 189), and who reminds us on that occasion of the long tradition of the conception of laws as inference rules.[21] From a purely logical point of view, it was always difficult to see the difference between the truth of $\bigwedge x(Px \rightarrow Qx)$ and the validity of the rule "for any a, infer Qa from Pa". However, I find that the aspect of persistence, which was so crucial for me, is more salient in the talk of inference rules. Thus, what appeared to be merely a metaphorical difference turns out to have a precise induction-theoretic basis. It should have been clear, in any case, that ranking functions are (possibly very complex) inference rules, indeed, as my analysis of normal conditions has shown, defeasible inference rules that are believed to be reliable, but not necessarily universally valid. Hence, my account may perhaps be used to underpin Lange's much more elaborated theory, and conversely his many applications to scientific practice may confer liveliness and plausibility on my account.

To put the point differently, one might say that the emphasis in my account of laws is on the single case. The mark of laws is not their universality, which breaks down with one counter-instance, but rather their operation in each single case, which is not impaired by exceptions. Here, I clearly join Cartwright (1989) and her repeated efforts to explain that we have to attend to capacities and their cooperation taking effect in the single case. Her objective capacities or powers thus correspond to my subjective reasons as embodied in a ranking function, a correspondence which is salient again in the comparison of Cartwright (2002, sect. 2) with my Sections 3 and 4. However, as I already said, I am content here with my subjective correlate and do not discuss its objectivization.

This is what separates me from Cartwright also according to the classification of Earman and Roberts (1999). They distinguish accounts that try to provide truth conditions for *ceteris paribus* laws from accounts that focus rather on their pragmatic, methodological, or epistemological role, and they place Cartwright in the first group, whereas my account clearly belongs to the second. Hence, I appear to be exempt from their criticism. However, though I agree with many of their descriptions, e.g., when they say that "a '*ceteris paribus* law' is an element of a 'work in progress' " (p. 466), I feel that pragmatics is treated by them, as by many others before them, as a kind of waste-basket category that consists of a morass of important phenomena defying clear theoretical description.

This feeling is reinforced by Earman, Roberts, and Smith (2002), who motivate their pragmatic or non-cognitivist turn in Section 4 by their finding in Section 3 that there is no solution to the "real trouble with CP-laws" that we have "no acceptable account of their semantics" and "no acceptable account of how they can be tested" (p. ##). In a way, the main purpose of this paper was to answer this challenge. To be sure, I did not provide a semantics in the sense of specifying truth conditions. But I gave an "epistemic semantics" in the sense of describing the doxastic role of unqualified as well as hedged laws, and I gave an account of how things having this role can be confirmed and disconfirmed. Of course, I did so on a fairly rudimentary formal level not immediately applicable to actual practice. But often, I find, the gist of the matter stands out more clearly when it is treated from a logical point of view.

ACKNOWLEDGMENTS

I am deeply indebted to Christopher von Bülow, Ludwig Fahrbach, Volker Halbach, Kevin Kelly, Manfred Kupffer, Arthur Merin, and Eric Olsson for a great lot of valuable comments. I have taken up many of them, but I fear I have dismissed the more important ones, which showed me how many issues would need to be clarified and substantiated, and which would thus require a much longer paper. I am also indebted to Ekkehard Thoman for advice in Latin.

Nancy Cartwright remarks, at the very end of the introduction of her book (1989), that my views on causation are closest to hers. The closeness, though, may not be easy to discover. This is the first paper for several years in which I continue on our peculiar harmony. I dedicate it to her.

NOTES

[1] $\bar{A}$ is the complement or the negation of A.

[2] I apologize for the double negation; after a while one gets used to it.

[3] In Spohn (1988) I still took the range to consist of arbitrary ordinal numbers. But the advantages of this generality did not make up for the complications.

[4] I believe that if epistemologists talk of justification and warrant, they should basically refer to this relation of A being a reason for B; cf. Spohn (2001). That's, however, a remark about a different context.

[5] A relevant argument is provided by the so-called problem of old evidence. The problem is that after having accepted the evidence it can no longer be confirmatory. However, this is so only on the basis of the first condition. According to the second condition, learning about A can never change what is confirmed by A, and hence the problem does not arise. This point, or its probabilistic analogue, is made by Joyce (1999, sect. 6.4) with the help of Popper measures.

[6] Note, by the way, that this would also hold for an infinite domain of quantification. Hence, for ranking theory there is no problem of null confirmation for universal generalizations which beset Carnap's inductive logic.

[7] I am using here a technical notion of positive instance: a is a positive instance of G iff G_a, i.e. $Pa \to Qa$, is true. If $Pa \land Qa$, a positive instance in the intuitive sense, would be learnt, the beliefs would change, of course (at least given our assumptions that nothing except G is believed in κ).

[8] "Resilient" might be an appropriate term as well, but I do not want to speculate whether this would be a use of "resilient" similar or different to the one introduced by Skyrms; cf., e.g., Skyrms (1980).

[9] Here, $G_{\neq a}$ stands for $\bigwedge x(x \neq a \to G_x)$. Note that we have $\kappa(\neg G \mid \neg G_a) = 0$ according to both the persistent and the shaky attitude, simply because $\neg G_a$ logically implies $\neg G$.

[10] In arriving at this conclusion, I am obviously catching up with Ramsey (1929) who states very early and very clearly: "Many sentences express cognitive attitudes without being propositions; and the difference between saying yes or no to them is not the difference between saying yes or no to a proposition" (pp. 135f.). "... laws are not either" [namely propositions] (p. 150). Rather: "The general belief consists in (a) A general enunciation, (b) A habit of singular belief" (p. 136).

[11] But see my account of causal explanation in terms of ranking functions in Spohn (1991).

[12] Some call this a *ceteris absentibus* condition. My Latin expert informs me, though, that "*ceteris absentibus*" usually means only "other men (and not women or non-human things) being absent".

[13] My Latin expert also tells me that there is not really a good translation of "other things being normal" into Latin.

[14] The more familiar case will be that the laws do not predict that a quality Q is present or absent, but rather that a magnitude assumes different values in a given object. From a logical point of view this does not make much of a difference. Let us stick here to the simplest case.

[15] For the following discussion see in particular Cartwright (1983, ch. 2 and 3).

[16] The superposition in the second sense could also be conceived of as the contraction of a superposition in the first sense by $\neg \bigvee x(Px \land P'x)$.

[17] An alternative way to remove the apparent conflict between G and G', which was envisaged by Cartwright (1983, pp. 57ff.), is to say that G and G' are not about the same Q. Rather, G is about Q-as-caused- by-P, and G' about Q-as-prevented-by-P'. In substance, though, the problem of superposition remains the same under this alternative.

[18] But see Spohn (1993).

[19] I am still jumping between sentences and the corresponding propositions as seems convenient to me.

[20] See, e.g., the rich results collected in the papers in Carnap, Jeffrey (1971) and Jeffrey (1980).

[21] The insight that the issues concerning laws fundamentally rest on the theory of induction rather than the theory of counterfactuals is more salient in Lange (2000) than in Lange (2002). However, the theory of induction takes a probabilistic turn in Lange (2000, ch. 4), a move about which I have already expressed my reservations.

REFERENCES

Carnap, R., and R. C. Jeffrey, eds: 1971, *Studies in Inductive Logic and Probability*, Vol. I, University of California Press, Berkeley.

Cartwright, N.: 1983, *How the Laws of Physics Lie*, Clarendon Press, Oxford.

Cartwright, N.: 1989, *Nature's Capacities and Their Measurement*, Clarendon Press, Oxford.

Cartwright, N.: 2002, 'In Favor of Laws that are not *Ceteris Paribus* After All', this issue, pp. 423–437.

de Finetti, B.: 1937, 'La Prévision: Ses Lois Logiques, Ses Sources Subjectives', *Annales de l'Institut Henri Poincaré* 7. Engl. translation 'Foresight: Its Logical Laws, Its Subjective Sources', in H. E. Kyburg jr., H. E. Smokler (eds), *Studies in Subjective Probability*, John Wiley & Sons, New York 1964, pp. 93–158.

Earman, J. and J. Roberts: 1999, '*Ceteris Paribus*, There is No Problem of Provisos', *Synthese* **118**, 439–478.

Earman, J., J. Roberts, and S. Smith.: 2002, '*Ceteris Paribus* Lost', this issue, pp. 281–301.

Gärdenfors, P.: 1981, 'An Epistemic Approach to Conditionals', *American Philosophical Quarterly* **18**, 203–211.

Gärdenfors, P.: 1988, *Knowledge in Flux*, MIT Press, Cambridge, MA.

Glymour, C.: 2002, 'A Semantics and Methodology for *Ceteris Paribus* Hypotheses', this issue, pp. 395–405.

Halpern, J. Y.: 2001, 'Conditional Plausibility Measures and Bayesian Networks', *Journal of AI Research* **14**, 359–389.

Hempel, C. G.: 1988, 'Provisoes: A Problem Concerning the Inferential Function of Scientific Theories', *Erkenntnis* **28**, 147–164.

Jeffrey, R. C., ed.: 1980, *Studies in Inductive Logic and Probability*, Vol. II, University of California Press, Berkeley.

Joyce, J. M.: 1999, *The Foundations of Causal Decision Theory*, Cambridge University Press, Cambridge.

Kelly, K.: 1999, 'Iterated Belief Revision, Reliability, and Inductive Amnesia', *Erkenntnis* **50**, 11–58.

Kelly, K.: 2002, 'A Close Shave with Realism: Ockham's Razor derived from Efficient Convergence' (forthcoming).

Lange, M.: 2000, *Natural Laws in Scientific Practice*, Oxford University Press, Oxford.

Lange, M.: 2002, 'Who's Afraid of Ceteris-Paribus Laws? Or: How I Learned to Stop Worrying and Love Them', this issue, pp. 407–422.

Lewis, D.: 1973, *Counterfactuals*, Blackwell, Oxford.

Nayak, A. C.: 1994, 'Iterated Belief Change Based on Epistemic Entrenchment', *Erkenntnis* **41**, 353–390.

Niiniluoto, I.: 1972, 'Inductive Systematization: Definition and a Critical Survey', *Synthese* **25**, 25–81.

Ramsey, F. P.: 1929, 'General Propositions and Causality', in D. H. Mellor (ed.), *Foundations. Essays in Philosophy, Logic, Mathematics and Economics*, Routledge & Kegan Paul, London 1978, pp. 133–151.

Rott, H.: 1991, *Reduktion und Revision. Aspekte des nichtmonotonen Theorienwandels*, Peter Lang, Frankfurt am Main.

Schurz, G.: 1995, 'Theories and Their Applications: a Case of Nonmonotonic Reasoning', in: W. Herfel et al. (eds), *Theories and Models in Scientific Processes*, Rodopi, Amsterdam, pp. 269–293.

Schurz, G.: 2002, '*Ceteris Paribus* Laws', this issue, pp. 351–372.

Silverberg, A.: 1996, 'Psychological Laws and Non-Monotonic Logic', *Erkenntnis* **44**, 199–224.

Skyrms, B.: 1980, *Causal Necessity*, Yale University Press, New Haven.

Spohn, W.: 1983, *Eine Theorie der Kausalität*, unpublished Habilitationsschrift, München.

Spohn, W.: 1988, 'Ordinal Conditional Functions. A Dynamic Theory of Epistemic States', in: W. L. Harper, B. Skyrms (eds), *Causation in Decision, Belief Change, and Statistics*, Vol. II, Kluwer, Dordrecht, pp. 105–134.

Spohn, W.: 1991, 'A Reason for Explanation: Explanations Provide Stable Reasons', in: W. Spohn, B. C. van Fraassen, and B. Skyrms (eds.), *Existence and Explanation*, Kluwer, Dordrecht, pp. 165–196.

Spohn, W.: 1993, 'Causal Laws are Objectifications of Inductive Schemes', in: J. Dubucs (ed.), *Philosophy of Probability*, Kluwer, Dordrecht, pp. 223–252.

Spohn, W.: 1997, 'Begründungen a priori – oder: ein frischer Blick auf Dispositionsprädikate', in W. Lenzen (ed.), *Das weite Spektrum der Analytischen Philosophie. Festschrift für Franz von Kutschera*, de Gruyter, Berlin, pp. 323–345.

Spohn, W.: 2000, 'Wo stehen wir heute mit dem Problem der Induktion?', in R. Enskat (ed.), *Erfahrung und Urteilskraft*, Königshausen & Naumann, Würzburg, pp. 151–164.

Spohn, W.: 2001, 'Vier Begründungsbegriffe', in T. Grundmann (ed.), *Erkenntnistheorie. Positionen zwischen Tradition und Gegenwart*, Mentis, Paderborn, pp. 33–52.

Fachbereich Philosophie
Universität Konstanz
D-78457 Konstanz
Germany

CLARK GLYMOUR

A SEMANTICS AND METHODOLOGY FOR *CETERIS PARIBUS* HYPOTHESES

ABSTRACT. Taking seriously the arguments of Earman, Roberts and Smith that *ceteris paribus* laws have no semantics and cannot be tested, I suggest that *ceteris paribus* claims have a kind of formal pragmatics, and that at least some of them can be verified or refuted in the limit.

1.

Many, perhaps most, of the results of scientific inquiry are generalizations towards which we have a peculiar attitude: although their apparent logical form is general, and we assert them and use them in prediction, in designing experiments, in deriving new relationships, and in rejecting contrary hypotheses, they have actual or possible (feasibly possible) counterexamples, we know they have counterexamples, and we cannot say precisely under what conditions counterexamples will not arise in the future. The claims hold in "normal" conditions, whatever those are, but not otherwise. Earman, Roberts and Smith (*this issue*; hereafter ERS) call such claims "putative *ceteris paribus* laws". Putative *ceteris paribus* laws occur, for example, in almost everything we know about cellular biology, in all of the causal claims of the social sciences, and throughout the medical sciences, ERS say three things about such claims: they are not laws, we do not understand what they say (they have no "semantics") and they cannot be tested. So far as I understand ERS, their arguments are: *ceteris paribus* generalizations are not laws because they are vaguely qualified by "normally;" they have no semantics because we have no general specification of the meaning of "normal", and they cannot be tested because there is no description of a possible detectible event, or sequence of events, with which they are inconsistent – counterexamples are, *ipso facto*, not normal.

Partially in response to these challenges, also in this issue, Spohn (*this volume*) offers a brilliant formal analysis of laws as pairings of general claims and ordinal rankings of propositions, rankings that represent dispositions, the disposition not to surrender the predictions of the generalization in the face of counterexamples. To test such laws, or at least to learn about

 Erkenntnis **57**: 395–405, 2002.
© 2002 *Kluwer Academic Publishers. Printed in the Netherlands.*

them, he suggests, is just to alter, by an appropriate procedure, the strength of one's dispositions to persist in trusting the predictions of a generalization as examples and counterexamples accrue. Along the way he gives an elegant, if perhaps unsatisfying, explanation of the "normal" conditions for a *ceteris paribus* law: the disjunction of all of the conditions in which the generalization holds.

I leave it to ERS whether Spohn's account responds to their arguments. I have a different take on the issues about belief and learning involved. I defer to Humpty Dumpty on whether *Ceteris paribus* generalizations, even those we believe, should be called "laws", but I think ERS do implicitly have a view of the pragmatics, if not the semantics, of such generalizations, and do implicitly have a view of testing or learning that can be made explicit. Whether it leads to their conclusions is another matter. In any case I shall foist it on them. When ERS say *ceteris paribus* claims have no semantics, I take them to mean that there is no intersubjective, general account of the sense of "normal"; that allows, however, for subjective accounts, for treating "normal" as a kind of indexical, and allows, if not a semantics, a pragmatics. When ERS say that *ceteris paribus* claims cannot be tested, I take them to be saying that there are is no possible finite sequence of observations that decisively refute or verify such claims; that allows, however, that such claims might be refuted or verified in the limit. I propose to investigate the howevers.

2.

I take the point of view that the aim of inquiry is first cognitive, and only derivatively practical. The aim of inquiry is to come to believe interesting truths and to avoid believing falsehoods (practical matters, the necessities of action, influence what is of interest, of course); the business of epistemology is to say what attitudes we should have towards hypotheses generated in the course of such inquiry; and the purpose of methodology is to provide the most efficient, feasible and reliable procedures to fulfill the aims of inquiry. Spohn takes the second of these as his starting point; I take the third. They interact, of course. Testing, a part of methodology, is the business of specifying procedures that will yield detectible outcomes that will, or should if we follow a related procedure, alter our attitude towards hypotheses and take us a step towards the aim of inquiry.

There is a philosophical framework, originally due to Hilary Putnam but nowadays most developed in computer science, for proving things about how methods can and cannot lead to the truth, reliably, efficiently and feasibly. It is quite abstract (none of the juicy stories here about how

Darwin discovered evolution or about Faraday and lines of force) but has the virtue that it can be applied in almost any context.

The set up is this: There is a set **H** of hypotheses, which may be infinite and which may not each be describable at some moment, but each of which can eventually be generated. Logically possible hypotheses not in **H** constitute "background knowledge" – what is known not to be true, or assumed not to be true in the context of inquiry. For each hypothesis h in **H** there is a set of possible data streams consistent with h. Each data stream is a sequence of objects, which may be descriptions or facts or events or situations or bangs on the sensory organs, but which are intended to represent the data obtained from experiment or observation. We allow that the same object may occur more than once in a data stream. I will assume that the data streams are unending, although with some technical reformulations that is inessential. The idea is that each element of a data stream is something that, if it occurs, an investigator following a method of inquiry detects. I will refer to a set of hypotheses and their associated data streams as a *discovery problem.*

In this setting a method of inquiry is simply a (partial) function whose domain is all initial segments of all data streams for all hypotheses in **H** and whose range is **H**. If the value of a method for a particular initial segment of data is h, I will say for brevity that the method *conjectures* h on that initial segment. The reliability, efficiency and feasibility of a method can have many forms. For feasibility, one can require that the partial function have any particular computational complexity one wishes; for efficiency one can require that a method get to the truth as fast as any method can in general, where speed can be measured, for example, by how often the method changes the hypothesis it conjectures.

My concern will be with reliability, which I will take to be this: A method is reliable for a discovery problem if and only if for every h in **H**, and for every data stream d, the value of the method (remember, a method is a function from initial segments of data streams to **H**) for all but a finite number of initial segments of d is h if and only if d is a data stream for h.

Why not require something stronger for reliability, for example that the method, which is a partial function, be correct on the first conjecture it makes – in other words, that the method remain silent until it has enough data to be sure which hypothesis in **H** is true, and then report that hypothesis, and always be right? The answer is that for general, universally quantified claims, no such method exists that decides reliably between the claim and its negation. The point is in *The Meno*, and the proof is in Sextus Empiricus. The sort of reliability I defined first, usually called limiting reliability in the literature, is the best we can do when we are talking, as we are here, of universally quantified claims.

The limiting reliability criterion just stated is equivalent, for most problems, to Bayesian convergence to the truth, provided no computational constraints are placed on the method. That is, there exists a reliable method for a discovery problem if and only if there are is a prior probability assignment to the hypotheses and initial segments of data streams such that, conditioning on the data, the probability distribution converges to 1 for the correct hypothesis. In some deterministic discovery problems this is a great advantage, because one can describe procedures for conjecturing hypotheses that accord with Bayesian convergence criteria without the technicalities of distribution theory or worries about whether the priors have been specified appropriately, all of which turn out to be gratuitous. When, more realistically, computational constraints are imposed on methods, the equivalence disappears and the learning theory framework dominates. Even if one requires only that methods be Turing computable: there are discovery problems that are solved reliably, as described above, by Turing computable methods, but for which convergence to the truth in all cases is not possible with Turing computable probability measures. The converse is not true.

Learning theory is part of methodology, and it says nothing about what attitude one should have towards the conjectures of a method, even a reliable method. Most of epistemology – Bayesian epistemology, Spohn's ranking theory, sense datum theory, classical skepticism, whatever – is, in one way or another, about that question. There are connections, of course. Most skeptical arguments, for example, turn on elementary considerations about the nonexistence of methods to reliably solve formally simple discovery problems, and ERS present us with the skeleton of a skeptical argument about *ceteris paribus* generalizations. In the other direction, reliability demands can be imposed on schemes for changing ranking functions (just as on schemes for changing probabilities), and Kelly (1998), has shown that only a few (one of which is due to Spohn) of the many proposals meet that requirement. And there are many variations on learning theory that more closely entangle epistemology and methodology. Instead of giving particular hypotheses as output, we could consider methods that give probability distributions over hypotheses (posterior probabilities are just one possibility), or partial orderings over hypotheses (Spohn's orderings are an example), or any of a variety of scores that are used in practical statistics when posterior probabilities are infeasible to calculate, or scores that are a function of the number of data points since the method changed its value, etc. We can equally consider the advantages and disadvantages of communities of inquirers using different methods for the same problem. That has been done for some while by computer scientists, by a few economists, and in philosophy, only by Philip Kitcher (1990).

Whatever the output of a method, the question arises as to its normative role in practical action, and only the Bayesians have a full, if highly idealized, story to tell: take the action with highest expected utility. (The recommendation is chiefly a philosophers' and economists' toy. I have never seen it done in any real scientific application, because even with explicit utilities it is computationally intractable in realistic cases unless one is very dogmatic about the possible hypotheses.) Endless alternatives are possible for forming propositional attitudes or making decisions using the output of a learning theoretic framework in which a single hypothesis is conjectured at a time, but I will advocate none of them here.

3.

Formal linguistics is not just semantics. It includes pragmatics as well, the analysis of speech acts whose truth conditions depend on the speaker, and more generally the analysis of indexical expressions whose denotations depend on time and place and context: *I, you, here, now.* Our understanding of language is not poorer, but richer, for recognizing and analyzing indexicals, even if we step out of formal "semantics" to something else.

Indexicals are possible in methodology as well. They are championed, in other terms, by philosophers who are *relativists*, who think that the truth value of a sentence in a language depends on something else besides the world and the language, on the endorsement of the community, or the conceptual scheme, or the method of inquiry, or something, and that in virtue of that something the truth values of sentences change in the course of inquiry. Sometimes this change is described as a change of language: the meaning of the sentence changes, depending on the something, while the syntax is kept, but I think the former way is easier to keep in mind without confusion.

Developments of relativist theories are usually placed within historical or sociological discussions, and their advocates are generally thought of as anti-formal. That is a pity, because formulated within learning theory they have an intricate structure, of interest to us here because it is a useful model for understanding *ceteris paribus* generalizations. Suppose we have a discovery problem that is ill-posed in the following respect: the set of data streams consistent with a hypothesis is itself a function of the conjectures of a method, or, put another way, for a given data stream, the hypothesis it corresponds to may change at any point in response to the conjecture

[123]

produced by a method. The formal details are developed in Kelly and Glymour (1992) and in Kelly (1996), and Kelly et al. (1992), but the idea should be clear enough.

The notion of limiting reliability of a method for a discovery problem still makes sense when relativized, in fact several distinct senses. In one sense of reliability, a method is reliable if there comes a time after which truth values cease to jump around and there also comes a time after which the method is always correct in its conjectures under the stable truth values. More interestingly, the truth values may change without end, but a reliable method eventually changes in twain and after a finite number of errors produces the correct hypothesis ever after. Problems of this kind have an interesting complexity, and if the allowable semantic jumps must be permutations of some finite set there is a universal method that will solve any solvable discovery problem.

The discovery of *ceteris paribus* claims is not relativistic inquiry, but put with a learning theoretic framework there is a similar indexicality and a structural similarity to the relevant notions of convergence to the truth.

4.

I will give a more general form to the third ERS claim that putative *ceteris paribus* laws are not testable: there is no procedure for discovering the truth or falsity of *ceteris paribus* claims, even in the limit. My justification is that if there were some procedure that reliably, efficiently and feasibly led us to the truth or falsity of *ceteris paribus* claims but did not involve "testing", the ERS claim would be, as Adolf Grunbaum likes to say, otiose, and if testing were not part of some procedure for getting to the truth, it would not serve the aims of inquiry.

I will interpret ERS as follows: A *ceteris paribus* **claim** has the form, *normally X*, or *ceteris paribus X*, where X contains no normalcy or *ceteris paribus* claim. An **instance** of *normally X* is a universal claim of the form *for all x, if A(x, z) then X (x)*, where A contains no normalcy claim but may have any quantifier structure. I will abbreviate this formulation as *if A then X. Ceteris paribus* generalizations, or claims that *normally, X*, are universal conditionals in which the antecedent is indexical and typically unexpressed; put another way, the apparent qualifier "normally" acts logically as a propositional function variable whose value is somehow determined in each case. *Ceteris paribus* generalizations are true for a case, a circumstance, a situation, if the appropriate value for the indexical antecedent for that case results in a true conditional, that is, if there is some appropriate instance of *normally X* that is true for the case. *Ceteris paribus* generalizations are true if they are true for every case. The issue is what determines which instances are "appropriate." I will pursue the idea that the learner – the inquirer – does the determining, although in all of what follows that determination could be made by any arbitrary community, e.g., the scientific community to which the learner belongs.

The speaker who endorses a *ceteris paribus* generalization endorses at least two things: there exist conditions A such that *if A then X*, and if the conditions he presently estimates to be such an A hold (his present normalcy conditions), then X holds of the case. The scientific or other interest of the *ceteris paribus* claim will of course turn on how frequently the normalcy conditions obtain, on how often they change, on how many logically independent propositions X share the same normalcy condition, and on other factors that are part of the routine of scientific assessment and debate.

The semantic oddity is that the *ceteris paribus* claim is indexical over the antecedent, that is, over the normalcy condition, but not over the consequent of the *ceteris paribus* claim, and, further, that the speaker, or inquirer, or learner, not the data, determine the normalcy condition for each case-but not its truth or falsity. In the course of inquiry into a *ceteris paribus* generalization the inquirer may alter his normalcy conditions for a proposition X – his current estimate of the antecedent to form the appropriate instance – while keeping the consequent, X, fixed, all the while endorsing the same tacit indexical conditional: *Normally, X.* (He may, of course, also abandon his endorsement of the *ceteris paribus* claim altogether.) In conjecturing *Normally, X*, the learner expresses a prediction about X and the conditions for normalcy that will obtain in the future. Learning "strict" laws is just the limiting case in which the normalcy condition is vacuous.

Less psychologically, and a bit more formally, a *ceteris paribus* learner for hypothesis X in the context of a particular discovery problem is a function that specifies, for each initial data segment, a consistent proposition A that (for simplicity) I will assume is bivalent for each initial data segment, and conjectures *if A then X* or alternatively, *if A then not X* for the next datum. (Recall that each datum may be a complex situation.) In the hands of the learner, the proposition *normally X* generates from the data a sequence of more specific conjectures: *if A1 then X, if A2 then X, ...*, *if An then ~X, ...* where some or all of the A_i and A_j may be the same proposition. The learner verifies that *normally X* for a data sequence if only a finite number of these conjectures is in error or is of the form *if An then ~X*, and falsifies that *normally X* for a data sequence if only a finite number of these conjectures is in error or is of the form *if An then X*. Reliable

verification or falsification is respectively verification or falsification over
all data streams in the discovery problem.

5.

The formal picture is coherent, and it provides both a semantics and a
methodology for *ceteris paribus* claims. *Normally X* is indexical because
the content of "normally" depends on the inquirer, but the ideal inquirer's
sequence of conjectures, together with the data, determine whether *normally X* is true in any case, whether *normally X* is true in general, and
whether normally X is true always after some time. Even so, one may
wonder if the original doubts, formulated in ERS claims, have been
addressed.

A second party, trying to determine whether a learner L can learn a
claim, *normally X*, seems to be faced with a problem if *L* cannot specify, a
priori, *L*'s function from finite data sequences to normalcy conditions for
X. That may be an intuition behind the ERS doubts about *ceteris paribus*
laws. They put the issues as semantic and methodological, but perhaps
they are really about others and not about ourselves: how are we to know
what someone means – what empirical conditions would make what is said
true or make it false – when someone says "normally" *X*? But it would
be a confusion to reject *ceteris paribus* laws on these grounds. Learning
the learner's function from data sequences to normalcy conditions for *X*
is much harder, unnecessarily harder, than learning whether the learner
converges to the correct truth value (or to the eventual truth value) about
"normally *X*." We can learn the latter in the limit (assuming, of course, that
in each case the truth or falsity of the current normalcy condition for *X* and
the truth or falsity of *X* can be determined). We simply conjecture "yes"
when the current normalcy criterion yields a correct instance of "normally,
X" and "no" otherwise.

The fact remains that some learner might, by having a policy of formulating stronger and stronger normalcy conditions, trivialize the content of
the claim that *normally X*. Nothing in either the semantics or epistemology
of *ceteris paribus* claims requires that a learner have such a policy, such
a normalcy function, and when we reasonably suspect someone does we
reasonably doubt his scientific seriousness. Epistemologically, the learning theoretic analysis of normalcy resembles in this respect the learning
theoretic analysis of truth that is relative to the learner's conjectures about
the truth. There are consistent relativistic systems in which the learner's
conjecture that *X* makes *X* true, but these are not all, or even the most
interesting, relativistic systems. Alternatively, of course, a learner might

be a fake who only pretends to a *ceteris paribus* hypothesis by observing whether *X* obtains in a case and also observing other features of the case and, if X obtains, announcing that conditions are normal, and, if *X* does not obtain, announcing a normality condition that he knows is not satisfied in the case. We have strategies for detecting such fakery, and they are imbedded in normal scientific practice.

6.

This preceding is my answer to the ERS objections to *ceteris paribus* laws, but it may be quite wrong. It may be that the real process of in-quiry into many *ceteris paribus* laws about *X* aims to discover a fixed (but possibly disjunctive) set of features of the normal world sufficient for *X* and preferably almost necessary for *X*, or perhaps for many hypotheses simultaneously. If so, such *ceteris paribus* generalizations are simply place holders in the search for the kind of laws that ERS call strict. *Ceteris paribus, X*, is what we say when we believe that there are strict condi-tions *A* such that for all cases, if *A* then *X*, and we are able to specify *A* incompletely and believe we can identify normal conditions when they are present, but are disposed to investigate the features of normalcy further if counterexamples to *X* occur, and aim eventually to eliminate all normalcy qualifiers. I do not think this is so, but it may be.

A more likely alternative is that our actual strategy is sometimes to use counterexamples to generate new hypothesis, new *X*s, that contain within them the conditions that apply when counterexamples occur. Per-haps *ceteris paribus* hypotheses are themselves part of a discovery strategy for new *ceteris paribus* hypotheses, and that is their special virtue. The genes of mouse liver cells express in a particular way on the surface of the Earth. Call the details *X*. Hammond discovered that in the microgravity aboard the Space Shuttle in Earth orbit, mouse kidney cells express their genes quite differently, call the details *Y*. Perhaps what we should now say is this: normally on Earth mouse kidney cells express as X, and normally in microgravity mouse kidney cells express as *Y*.

7.

So, I have more confidence that ERSs arguments are wrong than that my reconstruction of the semantics (ok, pragmatics) and methodology of *ceteris paribus* generalizations is right. In any case, the intuitions that I have guessed lie behind two of their three claims lead to an interesting analysis with interesting possibilities, if not to their conclusions, and I am grateful for their provocation. There is a later, positive proposal in their paper for which I am not grateful, and I cannot let it pass without comment.

Explicitly discussing causal claims, such as "smoking causes cancer" ERS say that this: they have no content, they say nothing, they merely "signal" or "express" the empirical data that prompts their assertion, and perhaps also signal a "research program" to "explain" their consequents in terms of their antecedents. Taking the proposal seriously, the introduction and discussion sections of most articles in *Science*, and all articles in *Cell*, should be excised, or at least retitled "Signaling." One can imagine the conversations:

> Russian scientist who knows only of Russian data on smoking and cancer:
>
> "Smoking causes cancer."
>
> An oncologist, who does not know of Russian data and has a different research program:
>
> "Smoking causes cancer."
>
> Russian scientist:
>
> "We agree, then."
>
> Earman, Roberts and Smith:
>
> "No you don't."

Meanwhile we can reflect on the predicament of the oncologist who, saying "smoking causes lung cancer", signals her program to "explain" lung cancer from smoking while the content of her explanations of lung cancer from smoking, beyond references to data, are nothing but more signals that she intends to carry out research to "explain" lung cancer from smoking. And on and on, forever.

In a time in which many first class minds are simultaneously struggling towards an understanding of causation and reliable methods for determining causal relations in complex circumstances, we should not be grateful for this.

REFERENCES

Kelly, K. T.: 1995, *The Logic of Reliable Inquiry*, Oxford University Press, Oxford.

Kelly, K.: 1998, 'Iterated Belief Revision, Reliability, and Inductive Amnesia', *Erkenntnis* **50**, 11–58.

Kelly, K. T. and C. Glymour: 1992, 'Inductive Inference from Theory Laden Data', *Journal of Philosophical Logic* **21**, 391–444.

Kelly, K., C. Juhl, and C. Glymour: 1994, 'Reliability, Realism, and Relativism', in P. Clark (ed.), *Reading Putnam*, Blackwell, London, pp. 98–161.

Kitcher, P.: 1990, 'The Division of Cognitive Labor', *Journal of Philosophy* **87**, 5–22.

Kitcher, P.: 1995, *Advancement of Science: Science without Legend, Objectivity without Illusions*, Oxford University Press, Oxford.

Clark Glymour
Department of Philosophy
135 Baker Hall
Carnegie Mellon University
Pittsburgh, PA 15213
U.S.A.
Email: cg09@andrew.cum.edu
Institute for Human and Machine Cognition,
University of West Florida

MARC LANGE

WHO'S AFRAID OF *CETERIS-PARIBUS* LAWS? OR: HOW I LEARNED TO STOP WORRYING AND LOVE THEM

ABSTRACT. *Ceteris-paribus* clauses are nothing to worry about; a *ceteris-paribus* qualifier is not poisonously indeterminate in meaning. *Ceteris-paribus* laws teach us that a law need not be associated straightforwardly with a regularity in the manner demanded by regularity analyses of law and analyses of laws as relations among universals. This lesson enables us to understand the sense in which the laws of nature would have been no different under various counterfactual suppositions – a feature even of those laws that involve no *ceteris-paribus* qualification and are actually associated with exceptionless regularities. *Ceteris-paribus* generalizations of an 'inexact science' qualify as *laws* of that science in virtue of their distinctive relation to counterfactuals: they form a set that is *stable* for the purposes of that field. (Though an accident may possess tremendous resilience under counterfactual suppositions, the laws are sharply distinguished from the accidents in that the laws are collectively as resilient as they could logically possibly be.) The stability of an inexact science's laws may involve their remaining reliable even under certain counterfactual suppositions *violating* fundamental laws of physics. The *ceteris-paribus* laws of an inexact science may thus possess a kind of necessity *lacking* in the fundamental laws of physics. A nomological explanation supplied by an inexact science would then be irreducible to an explanation of the same phenomenon at the level of fundamental physics. Island biogeography is used to illustrate how a special science could be autonomous in this manner.

1.

First things first. Defer the venerable problem of specifying the difference between claims expressing laws and claims expressing accidents. Consider just the *meaning* of a generalization qualified by "in the absence of disturbing factors", "other things being equal", "unless prevented", or something like that. The qualifier may simply list some factors and demand the absence of anything similar. For example:

> In the eighteenth-century British navy, only aristocrats were commissioned officers – unless the individual was the protégé of an aristocrat, or there was a case of fraud, or he distinguished himself in an especially gallant manner as a tar, or something like that.[1]

Erkenntnis **57**: 407–423, 2002.
© 2002 *Kluwer Academic Publishers. Printed in the Netherlands.*

Despite the qualifier's open-endedness, this claim is perfectly meaningful. It's easy to imagine what would constitute a counterexample to it and what would explain away an apparent exception. Let's elaborate this common idea (Scriven (1959), Molnar (1967), Rescher (1970), Hausman (1992)).

If a *ceteris-paribus* clause is meaningful, there is a tacit understanding of what makes a factor qualify as "disturbing" or as "like" the examples listed. Perhaps at the mere mention of any given factor, there would be immediate agreement on whether or not it qualifies as "disturbing". But this kind of unreflective unanimity is neither necessary nor sufficient for the *ceteris-paribus* clause to be meaningful. It's unnecessary because a consensus may develop only after the given factor is carefully compared to canonical examples of disturbing factors. (Or not even then.) There must, at any rate, be sufficient agreement on the relevant respects for comparison that analogies with canonical examples could supply a compelling reason for (or against) characterizing a given factor as "disturbing". Thus, when agents contemplate applying "All F's are G, *ceteris paribus*" to a given F, they can justify their belief that the *ceteris-paribus* condition holds (or doesn't), and their justification for deeming a given factor "disturbing" doesn't depend on their first ascertaining whether the given F is G.

Take the "law of definite proportions":

> Any chemical compound consists of elements in unvarying proportions by mass, *ceteris paribus*.

This qualifier could have been expressed as

> ... unless the compound is like ruby or like polyoxyethylene or something like that.

To understand this qualifier, one must know at least some of the following (Christie 1994). Ruby is composed of aluminum, oxygen, and chromium, different samples differing by even a factor of 5 in their chromium per unit mass of oxygen. Aluminum atoms are bonded to oxygen atoms, which are bonded to one another, forming a network running through the solid. Randomly, chromium atoms replace aluminum atoms. (They are similar in size and bonding capacities.) Ruby is $(Al, Cr)_2O_3$; its proportions are indefinite. Polyoxyethylene, in contrast, is a long-chain molecule beginning with CH_3, ending with CH_2OH, and containing many CH_2—O—CH_2's between. Because its length is variable, its proportions are indefinite; it is $CH_3(C_2H_4O)_nCH_2OH$.

Considering this background, one could offer compelling reasons for characterizing (say) olivine ($(Mg, Fe)_2SiO_4$) as *like* ruby and nylon as *like* polyoxyethylene. The *ceteris-paribus* clause has a determinate meaning. Likewise, indefinite proportions arising from the fact that different isotopes of the same element have different masses do *not* fall under *this* "*ceteris paribus*".

Seemingly, we could reformulate the *ceteris-paribus* clause as

unless the compound is a network solid or a polymer,

eliminating anything "vague". But the appearance of greater "explicitness" here is illusory. What is a network solid? It's something *like ruby* in the above respects. Once those respects have been grasped, no greater explicitness is achieved by replacing "like ruby" with "network solid".

Moreover, our new qualifier does not apply to exactly the same cases as our original one: not every case that is like ruby or like polyoxyethylene *or something like that* is clearly a network solid or a polymer. Take DNA. It fails to exhibit definite proportions since a DNA molecule's proportions depend on its length and its ratio of adenine-guanine to cytosine-thymine base-pairs. It isn't a network solid since it comes in discrete molecules. Yet DNA is like a network solid in one of its reasons for failing to exhibit definite proportions: certain subunits are able to replace others (of similar size and bonding capacities) randomly. (In DNA, however, those subunits are polyatomic.) Likewise, DNA is not a polymer, strictly speaking, since its "repeated" subunits are not all the same; there are two kinds, A-G's and C-T's, and there is no sequence they must follow. But DNA shares with polymers one reason for failing to exhibit definite proportions: a DNA molecule may be any length, and with greater length, a smaller fraction of its mass is contributed by endgroups.

A qualifier applying to exactly the same cases as the original one is

...unless the compound is a network solid or a polymer *or something like that.*

But this doesn't avoid "vagueness".

This doesn't show that there is no way to replace the original qualifier with something co-extensive yet "fully explicit". But in what sense would such an expression really *be* fully explicit? It would derive its content in just the way that the original qualifier did: by virtue of our implicit background understanding of what would count as compelling reasons for (or against) the correctness of applying it to a given case. There's nothing about how the "fully explicit" term *network solid* derives its meaning to distinguish it from the "vague terms"*ceteris paribus* and *like ruby* in the above examples.[2]

Earman and Roberts (1999) worry that if "in advance of testing" there is no statement of "what the content of a law is, without recourse to vague escape clauses", then there is no way to "guarantee that the tests are honest" because "the scientific community as a whole" could "capriciously and tacitly change what counts as an 'interfering factor' in order to accommodate the new data as they come in" (p. 451). But how does a claim *without* a "vague" *ceteris-paribus* clause supply the "guarantee" that Earman and Roberts crave? Even if the hypothesis is "explicit", there is never a *guarantee* that the scientific community will exercise good faith rather than tacitly re-interpret its hypothesis (and whatever statements of its meaning were issued in advance of testing it).[3] Suppose scientists originally make the definite proportions hypothesis "explicit" with the qualification

> ... unless the compound is a network solid or a polymer.

They then encounter DNA. There is nothing to *guarantee* that they won't unanimously, but incorrectly, say that DNA qualifies as a polymer in precisely the original sense.[4]

However, scientists *could* offer compelling reasons against the correctness of so classifying DNA: all of the canonical "polymers" so qualify because they involve many repeated (i.e., identical) small units, whereas DNA does not. This is *explicit enough* to resolve this case. The same kind of reasoning is available to determine a "vague" *ceteris-paribus* clause's applicability to a novel case. These examples are thereby distinguished from the twaddle that Earman and Roberts fear *ceteris-paribus* laws are in "danger" of becoming. For instance, suppose Jones says, "I can run a four-minute mile, *ceteris paribus*". He tries and fails. Were there no background for understanding the *ceteris-paribus* clause, there would be no basis for Jones to argue that it included "except on a muddy track". Suppose that Jones cashes out the qualifier as

> ... except on a muddy track, or when I have hurt my leg, or something like that.

Again he tries and fails to run a four-minute mile, and this time, the track is in good condition and he is healthy. Suppose Jones alleges that the race's having been held on the third Sunday in March is relevantly like the listed disturbing factors. Clearly, Jones now ascribes no determinate meaning to the qualifier.

Even when a given interpretation of a *ceteris-paribus* clause – such as "unless it involves many repeated (i.e., identical) small units" – is explicit enough for certain agents to determine the clause's applicability to one case, it may not be explicit enough for them to determine the clause's applicability to some other case. The ideas behind "identical" or "small" may require cashing out. This process will continue indefinitely as necessary, on a case-by-case basis. That should be considered "business as usual" rather than symptomatic of a poisonous vagueness.[5] Only the supposition that a "fully explicit", Platonic version of the definite proportions hypothesis exists would lead one to characterize such a *ceteris-paribus* clause as "lazy" (Earman and Roberts 1999, p. 461).[6]

2.

So *ceteris-paribus* clauses are nothing to worry about. But why should we love them?

One important lesson they teach us is that a law need not be associated straightforwardly with a regularity. It may be associated only with an inference rule that is 'reliable' – i.e., that leads to conclusions close enough to the truth for the intended purposes. To discover the law that all F's are G, *ceteris paribus*, scientists obviously must understand what factors qualify as 'disturbing'. But they needn't identify *all* of the factors that can keep an F from being G. They needn't know of factors that, when present, cause only negligible deviations from strict G-hood, or factors that, although capable of causing great departures from G-hood, arise with negligible frequency in the range of cases with which the scientists are concerned. Scientists need know only the factors that are non-negligible for the law's intended purposes: influences that arise sufficiently often, and can cause sufficiently great deviations from G-hood, that a policy of inferring F's to be G, regardless of whether they are under those influences, would not be good enough for the relevant purposes. Factors that may cause an F to depart from G-hood, but are negligible for the law's intended purposes, need not count as 'disturbing factors'. Hence, though it is a law that all F's are G, *ceteris paribus*, it is not true that all F's in the absence of disturbing factors are G.

When Boyle's law was discovered, for example, scientists must have understood its *ceteris-paribus* clause. But they did not know all of the factors that can cause gases to deviate from $PV = k$. They had not yet justified the kinetic-molecular theory of gases. They did not know that the forces exerted by gas molecules upon each other, the molecules' sizes, their adhesion to the container walls, the container's shape, and a host of other petty influences cause departures from $PV = k$. So in discovering that $PV = k$, *ceteris paribus*, scientists couldn't have discovered that $PV = k$ holds when the gas is 'ideal' in the above respects. Rather, the *ceteris-paribus* clause in Boyle's law covers the 'disturbing influences' recognized by scientists when discovering the law. It restricts the law's scope to relatively low pressures, high temperatures, and purposes tolerant of some inaccuracy. So qualified, $PV = k$ is good enough – *reliable*.[7] It "holds true for the 'permanent' gases under the experimental conditions usually employed in the common laboratory courses in physics, within the precision available in such experiments" (Loeb 1934, p. 140).

The *ceteris-paribus* clause has a pragmatic dimension, restricting the law's application to certain purposes. The F's aren't all (*ceteris paribus*) G. In both of these respects, *ceteris-paribus* laws aren't associated with regularities in the straightforward manner demanded by regularity analyses of law *and* analyses of laws as relations among universals.

This viewpoint is not new. Mill wrote:

> It may happen that the greater causes, those on which the principal part of the phenomena depends, are within the reach of observation and measurement .... But inasmuch as other, perhaps many other causes, separately insignificant in their effects, co-operate or conflict in many or in all cases with those greater causes, the effect, accordingly, presents more or less of aberration from what would be produced by the greater causes alone .... It is thus, for example, with the theory of the tides. No one doubts that Tidology ... is really a science. As much of the phenomena as depends on the attraction of the sun and moon ...may be foretold with certainty; and the far greater part of the phenomena depends on these causes. But circumstances of a local or casual nature, such as the configuration of the bottom of the ocean, the degree of confinement from shores, the direction of the wind, &c., influence in many or in all places the height and time of the tide .... General laws may be laid down respecting the tides; predictions may be founded on those laws, and the result will in the main ...correspond to the predictions. And this is, or ought to be meant by those who speak of sciences which are not *exact* sciences. (1961, 6.3.1, pp. 552–553)

One might cavil at honoring these *ceteris-paribus* generalizations with the exalted title "natural laws". (Mill (1961, 6.3.2, p. 554) said they "amount only to the lowest kind of empirical laws".) I could just shrug: what's in a name? But it would be better to clarify the reasons for regarding these *ceteris-paribus* generalizations as full-fledged laws. We must tackle the venerable question deferred at the outset: How do laws differ from accidents?

3.

Focus not on some dubious metaphysical picture of what laws have got to *be* to deserve the honor, but rather on what laws *do* in science. The received wisdom identifies several functions distinctively performed by laws, including supporting counterfactuals, grounding explanations, and being inductively projected from observed instances. Alas, none of these well-worn slogans suffices to pick out a role that no accident can play. Nevertheless, there is presumably some kernel of truth in these slogans, and they suggest that some *ceteris-paribus* generalizations can perform the roles characteristic of laws.[8] In a suitable context, Boyle's law supports the counterfactual, "Had the gas's pressure been half, its volume would have been double". The predictive accuracy of Boyle's law was confirmed by instances. We can use Boyle's law to explain why a certain gas's volume halved: we doubled the pressure on it.[9]

Let's get more specific about one of these roles. Laws supply reliable information on what the world would have been like had p been the case, for any counterfactual supposition p that is 'physically possible', i.e., logically consistent with every logical consequence of the laws.[10] In other words:

> *Nomic Preservation (NP)*: The laws would have been no different had p obtained, for any p logically consistent with every physical necessity.

For example, had I missed my bus this morning, the natural laws would have been no different: I would have been unable to get to my destination by making a wish and clicking my heels. Routinely, the laws are used to extrapolate what would have resulted from different initial conditions.

Counterfactual suppositions are often entertained in 'non-backtracking' contexts: in the closest p-worlds, the course of events remains just as it actually was until about the moment with which p is concerned, at which point history embarks on a different course (one that includes p). However, this change of course is disallowed by the actual laws if they are deterministic. (Let's suppose they are; this should make no difference to the laws' logical relation to counterfactuals.) Accordingly, David Lewis says that a small "miracle" (a violation of the actual laws) occurs in the closest p-world to make room for p to hold. But if laws must correspond to exceptionless regularities, then this "miracle" runs counter to *NP*. Lewis therefore rejects *NP*. But I cannot countenance "Had I missed my bus this morning, the laws of nature would have been different"!

Elsewhere (Lange 2000, pp. 73–76), I have examined various options for reconciling *NP* with the demands of non-backtracking, I've argued that the correct option is to reject the assumption that every law of a given possible world corresponds to an exceptionless regularity there. Rather, "All F's are G" can have exceptions in the closest p-world and still express a law there, so long as these violations fail to undermine the law's *reliability*. In the p-worlds that are optimally close in a non-backtracking context (where p is 'physically possible': consistent with the reliability of the actual laws), any violation of an actual law remains 'offstage': the departure from "All F's are G" is *negligible* because it occurs *before* the period of interest in the possible world's history. In a non-backtracking context, we don't care how p managed to come about, only what difference p would have made to subsequent events. For instance, in asking what the Earth would have been like today were there no moon (e.g., how the tides would have been different), we are not to consider how this moon-less state of affairs could have come about (e.g., whether the Earth would have been left with a stifling Venusian atmosphere had its CO_2 not been blown away in the cataclysmic impact creating the Moon).

This argument doesn't exploit the alleged peculiarities of *ceteris-paribus* clauses or special sciences. Even if there is a stratum of fundamental laws of physics corresponding to exceptionless regularities in the actual world, there must be an offstage exception to these laws in the closest p-world, although the laws are no different there. Such a law m is still *reliable* in the closest p-world (since any departure from it is offstage, in a non-backtracking context), and so "had p held, then m would have held" is close enough to the truth for the relevant purposes. This argument should lessen any urge we may feel to dig in our heels and say (in reply to the argument in Section 2) that since certain F's experiencing no 'disturbing factors' are nevertheless not G, it cannot be a law that all F's are G, *ceteris paribus*. The exceptions, like the offstage miracles, may be negligible for the intended purposes.

Let's turn this point around. In revealing that a law need not be true so long as it is reliable, *ceteris-paribus* laws point us toward the solution to a puzzle about all laws, even those without *ceteris-paribus* qualifiers: How do the actual laws remain laws under counterfactual suppositions considered in non-backtracking contexts? That's one reason I love *ceteris-paribus* laws.

4.

The best way to see that certain *ceteris-paribus* generalizations should be considered "laws" is by getting clearer on the laws' distinctive scientific roles and then observing that certain *ceteris-paribus* generalizations play those roles, especially in sciences like Mill's "Tidology". Let's pursue this strategy in connection with laws' capacity to support counterfactuals.

Is the range of counterfactual suppositions under which an accident would still have held *narrower* than the range under which a law would still have held? No. Suppose a large number of electrical wires, all made of copper, have been laid out on a table. Had copper been electrically insulating, then the wires on the table would have been useless for conducting electricity. Look what just happened: the law that all copper is electrically conductive obviously wouldn't still have held had copper been electrically insulating. But (in the envisioned conversational context) this counterfactual supposition fails to undermine the accident that all of the wires on the table are made of copper. So the range of counterfactual suppositions under which an accident is preserved can in some respects extend beyond the range under which a law is preserved.

Seemingly, then, there is no sharp distinction between laws and accidents in their power to support counterfactuals: some accidents are more fragile, other more resilient under certain sorts of counterfactual suppositions, and while laws are quite resilient, there is no sense in which they are more resilient than accidents. However, I think that adherents to this increasingly popular view are giving up too easily. A sharp distinction *can* be drawn here between physical necessities and accidents.

To begin with, many philosophers have endorsed something along the lines of *NP*: that the laws would still have held under any counterfactual supposition logically consistent with the laws. No accident is always preserved under all of these suppositions. But *NP* doesn't justify attributing to the physical necessities especially great counterfactual-supporting powers. That's because the *range* of counterfactual suppositions under consideration in *NP* has been designed expressly to suit the physical necessities. Suppose again that it is an accident that all of the wires on the table are copper. This accident's negation is physically possible, and so the accident is obviously not preserved under all physically possible suppositions. So it is *trivial* that no accident's range of invariance includes every counterfactual supposition logically consistent with the physical necessities.

What if we allow a set containing accidents to pick out a range of counterfactual suppositions especially convenient to itself: those suppositions logically consistent with every member of that set? Take a logically closed set of truths that includes the accident that all of the wires on the table are copper but omits the accident that all of the pears on my tree are ripe. Here's a counterfactual supposition consistent with every member of this set: had either some wire on the table *not* been made of copper or some pear on the tree *not* been ripe. What would the world then have been like? It is not the case (in many conversational contexts) that the generalization about the wires would still have held. (Indeed, in many contexts, it is the case for neither generalization that it would still have held.)

The same sort of argument could presumably be made regarding any logically closed set of truths that includes *some* accidents but not *all* of them. Given the opportunity to pick out the range of counterfactual suppositions convenient to itself, the set nevertheless isn't resilient under all of those suppositions. Trivially, every member of the set of *all* truths would still have held under any counterfactual supposition logically consistent with all of them, since *no* counterfactual supposition is so consistent.

Here, then, is my preliminary suggestion for the laws' distinctive relation to counterfactuals. Take a logically closed set of truths. (Truths – so as yet, I have left no room for *ceteris-paribus* laws. Stay tuned.) Take the counterfactual suppositions p that are logically consistent with every member of the set. Call the set *stable* exactly when for every member m of the set, m is reliable in the closest p-world(s): any departure there from m is negligible for the purposes for which this counterfactual world is being discussed (as I explained in connection with non-backtracking contexts, for example). So "had p held, then m would have held" is correct.

According to *NP*, the set of all physical necessities is stable. As I just argued, no set containing an accident is stable, except for the set of all truths, which is trivially so. What makes the physical necessities special is that *taken as a set*, they are resilient under as broad a range of counterfactual suppositions as they *could* logically possibly be: *all* of the physical necessities would still have held under *every* counterfactual supposition under which they *could all* still have held. No set containing an accidental truth can make that boast non-trivially (Lange 1999, 2000).

The logical necessities and the set of all truths are trivially stable. The set of physical necessities is stable non-trivially. Because it is as resilient as it could be, there is a sense of necessity corresponding to it. No sense of necessity corresponds to an accident, even one that would still have held under many counterfactual suppositions. The notion of 'stability' gives us a way out of the circle that results from specifying the physical necessities as the truths that would still have held under certain counterfactual suppositions: those consistent with the physical necessities.

How should this framework be applied to an "inexact science" like Mill's Tidology? What would it take for there to be laws of some such science – *ceteris-paribus* laws reflecting only "the greater causes"? We need to add two ingredients to our framework. First, we must permit a stable set to include not only truths, but also 'reliables' such as Boyle's law (for certain purposes).[11] Second, we must recognize that an inexact science's concerns are limited. A set is stable *for the purposes of a given inexact science* if and only if it is invariant under every counterfactual supposition of interest to the science and consistent with the set.

Take island biogeography, for example, which deals with the abundance, distribution, and evolution of species living on separated patches of habitat. It has been suggested that *ceteris paribus*, the equilibrium number

S of species of a given taxonomic group on an 'island' (as far as creatures of that group are concerned) increases exponentially with the island's area A: $S = cA^z$. The (positive-valued) constants c and z are specific to the taxonomic group and island group – Indonesian land birds or Antillean beetles. One theory (the "equilibrium theory of island biogeography" developed by Robert MacArthur and E. O. Wilson) purporting to explain this "area law" is roughly that a larger island tends to have larger available habitats for its species, so it can support larger populations of them, making chance extinctions less likely. Larger islands also present larger targets for stray creatures. Therefore, larger islands have larger immigration rates and lower extinction rates, and so tend to equilibrate at higher biodiversity. Nevertheless, a smaller island nearer the 'mainland' may have greater biodiversity than a larger island farther away. This factor is covered by the "*ceteris paribus*" qualifier to the "area law". Likewise, a smaller island with greater habitat heterogeneity may support greater biodiversity than a larger, more homogeneous island. This factor is also covered by "*ceteris paribus*". And there are others. Nevertheless, to discover the "area law", ecologists did not need to identify *every* factor that may cause deviations from $S = cA^z$, only the "greater causes". Like Boyle's law, the area law is intended to yield predictions good enough for certain sorts of applications, theoretical and practical, from planning nature reserves to serving as the first step in constructing divers ecological models.

Assume (for the sake of argument) that the "area law" is indeed accurate enough for these purposes (Figure 1). What must its range of invariance be for it to count as a law of island biogeography? There are counterfactual suppositions under which the fundamental laws of physics would still have held, but under which the "area law" is not preserved. For example, had Earth lacked a magnetic field, then cosmic rays would have bombarded all latitudes, which might well have prevented life from arising, in which case S would have been zero irrespective of A. Here's another counterfactual supposition: Had evolutionary history proceeded differently so that many species developed with the sorts of flight, orientation, and navigation capacities possessed by actual airplanes. (This supposition, albeit rather outlandish, is nevertheless consistent with the fundamental laws of physics since airplanes exist.) Under this supposition, the "area law" might not still have held, since an island's size as a target for stray creatures might then have made little difference to its immigration rate.[12] (Creatures without the elaborate organs could have hitched rides on those so equipped.)

Unlike the fundamental laws of physics, generalizations from inexact sciences aren't preserved under every counterfactual supposition consistent with the fundamental laws of physics. Accordingly, it has sometimes

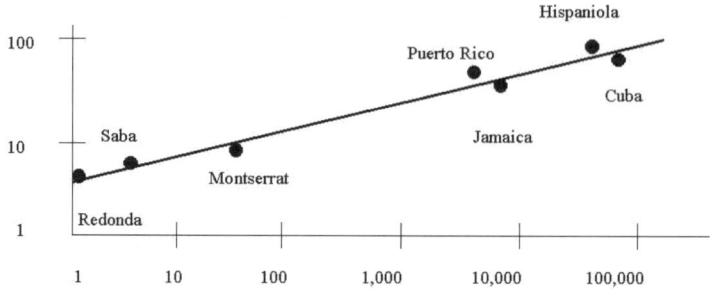

Figure 1. The area (in square miles) of various islands in the West Indies is depicted on
the *x*-axis. The number of amphibian and reptilian species on each island is depicted on
the *y*-axis (MacArthur 1972, p. 104).

been concluded that generalizations from inexact sciences fail to qualify
as natural laws. But this argument presupposes that a law *of* (say) *island
biogeography* would have to withstand the very same range of counterfactual perturbations as a fundamental law of physics. In an argument against
the possibility of laws in sciences other than fundamental physics, this
presuppositions amounts to begging the question.

The area law is not prevented from qualifying as an island-
biogeographical law – from belonging to a set that is stable for the purposes
of island biogeography – by its failure to be preserved under the two
counterfactual suppositions I just mentioned, although each is consistent
with the fundamental laws of physics. The supposition concerning Earth's
magnetic field falls outside of island biogeography's range of interests. It
twiddles with a parameter that island biogeography takes no notice of or,
at least, does not take as a variable. Of course, biogeographers draw on
geology, especially paleoclimatology and plate techtonics. Magnetic re-
versals are crucial evidence for continental drift. But this does not demand
that biogeography be concerned with how species would have been distributed had Earth's basic physical constitution been different. Biogeographers
are interested in how species would have been distributed had (say) Gon-
wanaland not broken up, and in how Montserrat's biodiversity would have
been affected had the island been (say) half as large. On the other hand,
biogeography is not responsible for determining how species would have
been distributed had Earth failed to have had the Moon knocked out of
it by a cataclysm early in its history. Biogeographers do not need to be
geophysicists.

The counterfactual supposition positing many species capable of covering long distances over unfamiliar terrain nearly as safely as short ones
over familiar territory is logically inconsistent with other generalizations
that would join the "area law" in forming a set stable for the purposes

of island biogeography. For example, the "distance law" says that *ceteris paribus*, islands farther from the mainland equilibrate at lower biodiversity levels. Underlying both the area and distance laws are various constraints: that creatures travel along continuous paths, that the difficulty of crossing a gap increases smoothly with its size (*ceteris paribus*), that creatures (apart from human beings) lack the organs, technology, culture, and background knowledge to make orientation and navigation virtually flawless, and so forth. These constraints must join the area and distance laws in the set stable for island-biogeographical purposes.[13]

The area law's *ceteris-paribus* clause does not need to rule out exceptions to these constraints. Although it isn't the case that the area law would still have held, had these constraints been violated, the area law's range of invariance may suffice for it to qualify as a law of island biogeography because *other* island-biogeographical laws, expressing these constraints, make violations of these constraints physically impossible (as far as island biogeography is concerned). Here's an analogy. Take the Lorentz force law: In magnetic field B, a point body with electric charge q and velocity v feels a magnetic force $F = (q/c)v \times B$. Presumably, it isn't the case that this law would still have held had bodies been able to be accelerated beyond c. But this law requires no proviso limiting its application to cases where bodies fail to be accelerated beyond c. That's not because there are actually no superluminal accelerations, since a law must hold not merely of the actual world, but also of certain possible worlds. The proviso is unnecessary because *other* laws of physics deem superluminal acceleration to be physically impossible. Hence, the Lorentz force law can belong to a stable set – can have the range of invariance demanded of a law of physics – without being preserved under counterfactual suppositions positing superluminal accelerations.

There may actually be *no* laws of island biogeography. Perhaps only a case-by-case approach makes approximately accurate predictions regarding island biodiversity. Perhaps there aren't just a few "greater causes", but many significant influences: weather and current patterns, the archipelago's arrangement, the island's shape, differences between island and mainland conditions, the character of mainland species, an island's habitat heterogeneity, different potential source areas and colonization capacities for different species, the presence on the island of competitors with and predators and parasites on potential colonists, and various idiosyncracies (such as the choices made by individual creatures and rare storms promoting immigration of species with low dispersal capacities). It's an open *scientific* question whether there are island-biogeographical laws. I want to understand what this issue *is about* – what it would take for such a science to have its own set of (*ceteris-paribus*) laws: to be *autonomous.*

A set stable for island-biogeographical purposes needn't include all of the fundamental laws of physics. The *gross* features of the laws of physics captured by constraints like those I've mentioned, along with the other island-biogeographical laws and the field's interests, may suffice *without the fundamental laws of physics* to limit the relevant range of counterfactual suppositions. For instance, had the geographic ranges of species been originally set and ever since maintained by miraculous Divine intervention, the "area law" might not still have held. But a set doesn't need to include the laws of physics in order to put this counterfactual supposition outside the range over which the set must be invariant in order to qualify as stable for island-biogeographical purposes. This supposition already falls outside that range in virtue of falling outside island-biogeographical concerns, which are limited to the *evolutionary* tendencies of island species. (Likewise for the supposition of deliberate human intervention.)

Similarly, the area law would still have held had there been birds equipped with organs weakening gravity's pull somewhat, assisting in takeoffs. The factors affecting species dispersal would have been unchanged: smaller islands would still have presented smaller targets to off-course birds and so accumulated fewer strays, *ceteris paribus*. The island-biogeographical laws's range of stability may thus in places extend beyond the range of stability of the fundamental laws of physics; the island-biogeographical laws don't reflect every *detail* of the physical laws.

The island-biogeographical laws's necessity derives from their range of stability. But that range is not wholly contained within the fundamental physical laws's range of stability (since it includes some suppositions inconsistent with the physical laws). Consequently, the physical laws's stability is not responsible for the island-biogeographical laws's stability. In other words, the island-biogeographical laws do not inherit their *necessity* from the fundamental laws of physics. The island-biogeographical laws's necessity is not borne by the fundamental physical laws; the range of stability of the island-biogeographical laws extends in some respects beyond (though, in other respects, is more limited than) that of the fundamental physical laws. The approximate *truth* of island-biogeographical laws might follow from the laws of physics and certain accidents of physics. But the *lawhood* of island-biogeographical laws – their stability (for the field's purposes) – cannot so follow. For that stability depends on their remaining reliable under certain counterfactual suppositions violating fundamental physical laws.

Hence, if there turned out to be island-biogeographical laws, island biogeography would have an important kind of autonomy. Because the lawhood of island-biogeographical laws would be irreducible to the lawhood of the fundamental laws of physics (and initial conditions), the nomological explanations supplied by island biogeography (of, for instance, Mauritius's biodiversity) would be irreducible to the explanations of the same phenomena at the level of fundamental physics (Lange 2000, Chap. 8).

That's pretty cool. It's another reason to love *ceteris-paribus* laws.

NOTES

[1] See Rescher (1970, p. 173).

[2] Of course, 'network solid' has a context-independent meaning whereas different tokens of '*ceteris paribus*' (and 'like ruby') may mean different things depending on the generalizations to which they are attached.

[3] Consider the arguments that in December 2000, the Florida Supreme Court changed rather than re-interpreted election laws.

[4] Thus, unreflective unanimity doesn't ensure that '*ceteris paribus*' has determinate meaning, just as Earman and Roberts (1999, p. 451) say.

[5] There may even be disagreement over whether some expression is explicit enough for the relevant agents to be able to give *reasons* for or against its application to various cases. For example, while five U.S. Supreme Court justices held in December 2000 that the 'intent of the voter' standard is not sufficiently explicit for ballot counters, Justice Stevens dissented, contending that it is no less vague than the customary 'beyond a reasonable doubt'.

Wittgenstein's rule-following point, which motivates my argument (Lange 1993, 2000), applies to *any* meaningful remark (descriptive claim, rule, whatever) – contrary to Earman and Roberts (1999, pp. 449–450).

[6] Earman and Roberts distinguish this "lazy" sense from the "improper" sense of '*ceteris paribus*' allegedly attached to (say) Coulomb's law. They see this law as requiring no qualification because it relates two bodies' charges and separation to the *component* electric forces they exert upon each other (1999, p. 461). Earman and Roberts

do not understand how anything short of a blanket anti-realism can motivate the notion that [a] component of a total impressed force is unreal. ... [M]odern physical theory from Newton onward gives two reasons to take certain component forces as having real ontological significance: first, the theory gives an account of how the component force arises from the distribution of sources (masses for the gravitational force, charges for the electrical force, etc.); and it promotes a form of explanation in which the total resultant force is obtained as a vector sum of the component forces that are due to sources. (p. 474)

Apparently, they argue that component forces are real because they are causal actors: the local causes of the net acceleration of the body feeling them, the effects ultimately of distant electric charges etc. Presumably, the picture endorsed by Earman and Roberts says that for each real component electric force acting on a body, there is a distant charge whose electric field causes exactly that component force. But this view leads to a problem. Classically, reality is ascribed to the electric field E (i.e., the net = total = resultant field) in virtue of its possessing energy with a density proportional to E^2. (See Lange 2002, Chap. 5.) But the individual electric fields E_i of various bodies cannot themselves each possess energy with a density proportional to E_i^2, else there would be the wrong total quantity of field energy. (That's because $[E_1 + E_2 + \cdots]^2$ doesn't generally equal $E_1^2 + E_2^2 + \cdots$.) Thus, the classical argument for the net field's reality doesn't carry over to its components' reality. So there's reason short of blanket anti-realism for interpreting the component forces (associated with these component fields) as unreal. (Also see Lange 2000, pp. 164–165.)

I've *not* followed Cartwright and Giere in characterizing as a ceteris-paribus law a claim (e.g., Coulomb's law) purportedly describing the action of a single influence. I've argued (Lange 2000, pp. 180–183) that a law like Coulomb's requires no provisos ruling out the presence of other forces. I *have* characterized as *ceteris-paribus* laws various claims (such as Boyle's law) that aim to characterize the *net* outcome of *all* the (non-negligible) influences. I had regarded the 'law of thermal expansion' as a law of the latter kind. If it is actually a law of the former kind (as Earman, Roberts, and Smith say), then it is not an example that serves my purposes.

[7] Compare Earman and Roberts (1999, p. 463).

[8] Compare Rescher (1970, pp. 170–171). See my (2000).

[9] A general account of confirmation and explanation would identify what it is about Boyle's law (its relation to counterfactuals? to unification? to causal powers?) that makes it explanatory.

[10] I restrict myself throughout to counterfactual suppositions (and logical consequences of the laws) that do not include expressions like 'law' and 'accident'. For details relevant to this section and the next, see my (2000).

[11] Accordingly, we must expand the notions of a 'logically closed set' and p's being 'consistent with every member of the set'. For example, p's consistency with m requires only that p be consistent with m's *reliability*: the claims to which we would be entitled, by reasoning from p in accordance with the inference rule associated with m, could all be close enough to the truth for the relevant purposes. Also, I'm assuming that each of the set's members is of interest to the field, so that its reliability for the field's purposes is non-trivial. See my (2000, Chap. 8).

[12] Suppose the 'target effect' to be so significant that without it, the species-area relation would have violated $S = cA^z$.

[13] See MacArthur (1972, pp. 59–60) on 'continuity principles'.

REFERENCES

Christie, M.: 1994, 'Philosophers versus Chemists Concerning "Laws of Nature" ', *Studies in the History and Philosophy of Science* **25**, 613–629.

Earman, J. and Roberts, J.: 1999, '*Ceteris Paribus*, There is no Problem of Provisos', *Synthese* **118**, 439–478.

Hausman, D.: 1992, *The Inexact and Separate Science of Economics*, Cambridge University Press, Cambridge.

Lange, M.: 1993, 'Natural Laws and the Problem of Provisos', *Erkenntnis* **38**, 233–248.

Lange, M.: 1999, 'Laws, Counterfactuals, Stability, and Degrees of Lawhood', *Philosophy of Science* **66**, 243–267.

Lange, M.: 2000, *Natural Laws in Scientific Practice*, Oxford University Press, New York.

Lange, M.: 2002, *An Introduction to the Philosophy of Physics*, Blackwell, MA.

Loeb, L.: 1934, *The Kinetic Theory of Gases*, McGraw Hill, New York.

MacArthur, R.: 1972, *Geographic Ecology*, Princeton University Press, Princeton.

Mill, J.S.: 1961, *A System of Logic*, Longmans Green, London.

Molnar, G.: 1967, 'Defeasible Propositions', *Australasian Journal of Philosophy* **45**, 185–197.

Rescher, N.: 1970, *Scientific Explanation*, Free Press, New York.

Scriven, M.: 1959, 'Truisms as the Grounds for Historical Explanations', in P. Gardner (ed.), *Theories of History*, Free Press, New York, pp. 443–475.

University of Washington
Box 353350
Seattle, WA 98195-3350
U.S.A.
mlange@u.washington.edu

NANCY CARTWRIGHT

IN FAVOR OF LAWS THAT ARE NOT *CETERIS PARIBUS* AFTER ALL*

ABSTRACT. Opponents of *ceteris paribus* laws are apt to complain that the laws are vague and untestable. Indeed, claims to this effect are made by Earman, Roberts and Smith in this volume. I argue that these kinds of claims rely on too narrow a view about what kinds of concepts we can and do regularly use in successful sciences and on too optimistic a view about the extent of application of even our most successful non-*ceteris paribus* laws. When it comes to testing, we test *ceteris paribus* laws *in exactly the same way* that we test laws without the *ceteris paribus* antecedent. But at least when the *ceteris paribus* antecedent is there we have an explicit acknowledgment of important procedures we must take in the design of the experiments – i.e., procedures to control for "all interferences", even those we cannot identify under the concepts of any known theory.

1. INTRODUCTION

I am generally taken to be an advocate of *ceteris paribus* laws throughout the sciences, even in physics. But what are *ceteris paribus* laws? According to John Earman, John Roberts, and Sheldon Smith (*this issue*), the distinctive feature of *ceteris paribus* laws is that they do not entail any strict or statistical regularities in the course of events. Nor do they entail any predictions, categorical or probabilistic. Earman, Roberts, and Smith also suppose that a *ceteris paribus* law is not explicit about what precise conditions have to obtain for the regularity after the *ceteris paribus* clause to hold; alternatively that the *ceteris paribus* clause is vague and cannot be stated in a precise form or a precise and closed form.

If that's what it takes, then what I have defended are not *ceteris paribus* laws. [1] The laws I talk about either can be stated in precise and closed form or they entail strict or statistical regularities in the course of events or both. The matter hinges, of course, on what we take to constitute a "precise and closed" description. This returns us to the old issue of how we should police the language of science. I suspect that I am far less strict about what is admissible as a description than are Earman, Roberts, and Smith. My reason is that I find that the less restrictive language is the kind of language that is regularly employed in exact science; and that attempts

Erkenntnis **57**: 425–439, 2002.
© 2002 *Kluwer Academic Publishers. Printed in the Netherlands.*

to reconstruct this language away produce scientific claims that are at odds both with the ways we test our scientific theories and with the ways we put them to use.

There are two kinds of formulation that I use to reconstruct scientific laws that I think Earman, Roberts, and Smith would reject, the first because of the language it uses, the second because of the limitations it supposes on the descriptive power of theory. I shall discuss each in turn in Sections 2. and 3. In Section 4. I shall take up the issue of testing. What I say about testing will not only defend the kinds of laws I discuss but also *ceteris paribus* laws as more generally conceived. Section 5. answers some criticisms that Earman et al., make in this volume against a connection I trace between induction and capacity.

2. POWERS VS. *ceteris paribus* LAWS

The language I use in reconstructing a number of scientific laws in the exact sciences (most notably physics and economics) is the language of *powers*, *capacities* or *natures* and related concepts such as *interfere*, *inhibit*, *facilitate*, and *trigger*. Those of us raised in the joint shadow of the Vienna Circle and British Empiricism were taught that these kinds of concepts must not appear in science. I was puzzled about this from early on since it seemed to me that many of the most important concepts I learned in physics are power or capacity concepts, *force* being the first, simple example.

A big obstacle to debate here is the problem of characterization: what criteria distinguish capacity concepts from OK concepts? Surely we do not want to adopt the characterization of the early British Empiricists that OK concepts are those built out of ideas that are copied from impressions. Operationalization was on offer for a while, but it seems to cut too narrowly since it rules out many central theoretical concepts. Nor do I think we can be content with Carnap's similarity circles and the *Aufbau*.

In my own early attempts to understand these empiricist strictures, I proceeded differently, in a way that generally works best for 'trouser' words (that is, for concepts whose primary meaning comes from what they rule out): figure out what is supposed to be wrong with the illicit concepts; the OK ones are those that don't have those problems. What then is supposed to be wrong with power concepts? One central worry comes from the fact that power concepts seem to be tied either too closely or too loosely to their related effects. This in turn is thought to lead to problems in testing claims about powers. I shall discuss these latter in Section 4.

That powers and their effects are tied too closely was the complaint of the old Mechanical Philosophers against Scholastic concepts (cf. Glanville 1661). What makes heavy bodies fall? Gravity. What is gravity? That which makes heavy bodies fall. For those like Ernst Mach (1893), who wished to provide explicit measurement procedures for the concepts of physics, Newton's second law seems to suffer the same difficulty. $F = ma$. What is it for a body of mass m to be subject to a total force of size F? A mass of size m is subject to force F iff its acceleration is a.

On the other hand, when the power is not defined in terms of the occurrence of its effects, there seems to be no fixed connection between the existence of the power and the occurrence of its effects, neither strict (i.e., universal) nor statistical. Aspirins have the power to relieve headaches; that is surely consistent with the fact that they do not always do so and perhaps there is no fixed statistical relation either. This objection to powers echoes an objection that Earman, Roberts, and Smith make to *ceteris paribus* laws.

There is a familiar way to fix this problem: insist that the effect is there after all whenever the power is.[2] One well known case of this arises in discussions of the problem of evil. God is omnipotent: He has the power to create any kind of world at all. Couple this with the auxiliary assumption that He is all good. The effect to expect is a benign world, full of delights. Instead we see plagues and poverty and vice. One stock response is that the world is all good despite appearances. We simply fail to see or perhaps to understand the situation properly. I take it that this kind of claim must be judged unacceptable by standards employed in successful science. The world does not appear good; it does not pass any of the standard tests for being good; and its effects are not the effects we are entitled to predict from a world of virtue and perfection.

Or consider Freudian claims, which we know distressed many followers of the Vienna Circle. Consider a crude version of one Freudian example. Freud maintained that the childhood experiences that the Ratman sustained have the power to make one desire the death of one's father. Freud says: "... he [the Ratman] was quite certain that his father's death could never have been an object of his desire but only of his fear. ... According to psychoanalytic theory, I [Freud] told him, every fear corresponds to a former wish which was now repressed; we were therefore obliged to believe the exact contrary of what he had asserted. ... He wondered how he could possibly have had such a wish, considering that he loved his father more than any one else in the world" (Freud 1909, p. 39). The Ratman did not recognize the desire for his father's death in himself, he appeared to others as a loyal and loving son and he had behaved just like someone concerned to ensure the welfare and safety of his father. But that's alright on Freud's view. The desire is really there; it is just unconscious and thus does not manifest itself in the usual ways.

Turn now to what Earman, Roberts, and Smith call "special force laws", like the law of universal gravitation (A system of mass M exerts a force of size GMm/r^2 on another system of mass m a distance r away) or Coulomb's law (A system with charge q_1 exerts a force of size $\varepsilon_0 q_1 q_2/r^2$ on another system of charge q_2 a distance r away).[3] These are not strict regularities. Any system that is both massive and charged presents a counterexample. Special forces behave in this respect just like powers. This is reflected in the language we use to present these laws: one mass *attracts* another; two negative charges *repel* each other. *Attraction* and *repulsion* are not among what Ryle (1949) called 'success' verbs. Their truth conditions do not demand success: X can truly attract Y despite the fact that Y is not moved towards X.

But perhaps, as with the delights of our universe or the Ratman's desire for the death of his father, the requisite effects are really there after all. Earman, Roberts, and Smith feel that the arguments against this position are not compelling. I think they are: the force of size GMm/r^2 does not appear to be there; it is not what standard measurements generally reveal; and the effects we are entitled to expect – principally an acceleration in a system of mass m a distance r away of size GM/r^2 – are not there either.

Contrast a different case (see Cartwright 1989, Ch. 4). In simultaneous equations models in economics each equation is the analogue of a special force law: each describes the operation of a single cause. When more than one cause is present, all the equations must be satisfied at once. So the pattern of behavior that occurs is one consistent with each equation separately. Unlike mechanics, the 'special force laws' in economics really are strict regularities (if true at all). The effects demanded by each law separately are really there – and they meet standard requirements for doing so: the effects of each appear to be there; standard measurements reveal them; and the effects of these effects are the ones we are entitled to expect.

The price level in economics is a contrast. It is calculated by summing the 'contributions' of a variety of different causes. But we do not want to think of the price level as literally composed of a lot of distinguishable parts as a wall is composed of its stones – the level from w to x is that due to the stock of money; from x to y, that due to the velocity of money; etc. This seems a highly unnatural reading to me. And it seems even more so when we move to engineering examples – say the construction of complex machines from simple ones or of circuits from combinations of resistors, capacitors and impedances – where the rules for how to calculate what happens when the parts act together are not by simple addition as they are in the case of *force* or *price level*.

Kevin Hoover in his extended study *Causality in Macroeconomics* backs up my point by criticizing the assumption of linearity of causal influences; that is, the assumption that "the influence of Y can be added to the influence of M and so forth" (Hoover 2001, p. 55) in calculating their effect when operating jointly. (In Hoover's example, Y is income, M, money and the effect is the interest rate.) Hoover complains, "But linearity is unduly restrictive" (*ibid.*, p. 55). Hoover is particularly concerned with the non- linearities arising from rational expectations theory, which imply that the influences of macroeconomic causes cannot be calculated just by addition. He illustrates with a mechanical example:

A gear that forms a part of the differential in a car transmission may have the capacity to transmit rotary motion from one axis to another perpendicular to it... The capacity of the differential to transmit the rotation of the engine to the rotation of the wheels at possibly different speeds is a consequence of the capacities of the gear and other parts of the differential. The organization of the differential cannot be represented as an adding up of influences nor is the manner in which the gear manifests its capacity in the context of the differential necessarily the same as the manner in which it manifests it in the drill press or in some other machine ... (*ibid.*, p. 55f)

Does all this matter? Pretty clearly it does not matter to the economics or to the physics under discussion. But it does matter to the metaphysics, particularly to the topic under discussion here – *ceteris paribus* laws. To see why let me explain how I see the difference between physics and many of the human sciences.

We study capacities throughout the sciences. Many of the central principles we learn are principles that ascribe specific capacities to specific features that we can independently identify, from the capacity of a massive object to attract other masses to the capacity of maltreatment of a child to cause that child to maltreat its own children, or, to mention the example that Earman, Roberts, and Smith discuss in their paper in this volume, the capacity of smoking to cause lung cancer.

What is special about physics then? Not that it does not offer knowledge about powers or capacities but rather that it has been able to establish other kinds of knowledge as well, knowledge that we can couple with our knowledge of capacities to make exact predictions. This additional knowledge is primarily of two kinds: (1) We know for the powers of physics when they will be exercised (e.g., a massive object *always* attracts other masses); and (2) we have rules for how to calculate what happens when different capacities operate together (e.g., the vector addition law for forces). This kind of knowledge is missing for many other subjects. That is why they cannot make exact predictions.[4]

Now for *ceteris paribus* laws. Consider the example of Earman, Roberts, and Smith (*this issue*)

(S)   CP, smoking causes lung cancer

of which they say, "If an oncologist claims that (S) is a law, then, we maintain, there is no proposition that she could be expressing (except for the vacuous proposition that if someone smokes, their smoking will cause them to get cancer, unless it doesn't) ..." I disagree. I take it that the proposition she is expressing is

(S′)   Smoking has the capacity to cause lung cancer.

a claim exactly analogous to the special force laws of physics. This is a precise claim: it states a matter of fact that is either true or not; it is not vague; and it has no *ceteris paribus* clause that needs filling in. So it does not suffer from those faults Earman, Roberts, and Smith ascribe to *ceteris paribus* laws. More central to their objections, it is testable, it makes predictions, and it entails regularities in the course of events, in this case statistical regularities. This is the topic of Section 4.

3. THE LIMITS OF SCIENTIFIC LANGUAGES

In the discussion so far I have been more liberal in my reconstruction of the language of science than most modern empiricists. I allow it to cover more, to talk about powers and capacities. Now I shall propose a way in which I think the languages of the different sciences can describe less.[5]

Consider Newton's second law, $F = ma$. What does it say? Many, probably including Earman, Roberts, and Smith, take it to describe a strict regularity. I think that it does so only conditionally. The claim we are entitled to believe from the vast evidence in its favor is this: *if nothing that affects the motion operates that cannot be represented as a force*, then. ... The two views collapse together if all causes of motion can be represented as forces. Why do I think many might not be?

Newtonian mechanics, like many other theories in physics, has, I believe, very much the structure that C. G. Hempel (1966) taught us theories have. It consists of internal principles, such as Newton's three laws of motion, which give relations among the central concepts of the theory, and bridge principles, which constrain how some of the concepts of the theory are applied. Many early logical positivists hoped that the bridge principles would lay out direct measurement procedures for all the concepts of the theory. They had to settle for less. The bridge principles match some theoretical concepts with concepts 'antecedently understood'.

In the case of Newtonian mechanics the primary bridge principles are the special force laws. These license a particular theoretical description – e.g. '... is subject to a force $F = GMm/r^2$' or '... is subject to a force $F = \varepsilon_0 q_1 q_2 / r^2$' – given a description in the vocabulary of masses, distances, times and charges – e.g., '... is a mass m located at distance r from a mass M', or '... is a charge q_1 located at distance r from charge q_2'.[6] Bridge principles provide strong constraints. The theoretical descriptions – in our example the individual force functions from the special force laws – are allowed *only if* the corresponding descriptions in 'antecedently understood' terms are satisfied.[7] (For example, "The force on a mass of size m is GMm/r^2 if and *only if* m is a distance r from a body of mass M".) The same thing is true of quantum mechanics and its bridge principles as well as quantum field theory, quantum electrodynamics, classical electromagnetic theory and statistical mechanics.

It is because of the issue of evidence that I urge that the bridge principles of these theories are so strongly constraining. I have looked at scores of applications and tests of the theories in my list, applications and tests of the kind that we take to argue strongly for the truth of these theories. In these cases the theoretical terms that have bridge principles are invariably applied via the bridge principles. This interpretation of the demands of bridge principles in turn puts a strong constraint on the descriptive power of the theory. Force functions can be legitimately applied only to situations that are described in bridge principles. Similarly for quantum Hamiltonians, classic electric and magnetic field vectors, and so on.

Can all causes of motion be correctly described using just the descriptions that appear in the bridge principles of Newtonian mechanics? To all appearances, not. We have millions of examples of motions that we do not know how to describe in this way. Consider one case where we eventually were successful. For centuries we knew about electricity and magnetism: e.g., rubbing a glass rod against cat fur can cause human hair to move; loadstones can move iron filings; and so forth. But we could not add these in as forces in Newton's second law. Eventually we evolved the formal, precise concepts of electric and magnetic charge as well as the bridge principles that assign them force functions. In so doing we came to ring-fence a host of macroscopic situations from all the rest of those that could cause motion but that we could not describe in Newtonian theory: *these* are ones involving attraction or repulsion between electrically or magnetically charged objects. But what of the vast remainder?

We have succeeded in applying Newton's second law to a vast, vast number of cases – but always of the same kinds: the ones that appear in our bridge principles. And there are still not many bridge principles included in Newtonian mechanics, even after 300 years.[8] We are not constantly expanding the theory, regularly producing new bridge principles to meet either new cases or the old familiar ones. Nor do we have continuous success in bringing these cases in under the old bridge principles. This suggests that these cases may well not fall under any descriptions for which there are force functions. And a handful of striking successes does not discount this worry.

My conclusion from these kinds of considerations is that we need to add to the basic 'equations of motion', like $F = ma$ or Schroedinger's equation, a special constraining condition: the equation holds so long as everything that can affect the targeted effect is describable in the theory. This is the formulation of the law that we have strong evidence for. Notice that it is in 'precise and closed form' and hence does not look like a *ceteris paribus* law on one of the criteria of Earman, Roberts, and Smith. But how do we test it?

4. TESTING

How do we test my versions of Newton's second law or the Schroedinger equation? In *exactly the same way* that we would test them if they had no condition attached. The same is true *mutatis mutandis* for capacity ascriptions and for certain kinds of more conventionally rendered *ceteris paribus* laws. Although there are important differences, let us for the sake of brevity lump all these together and consider them to be of the form, 'If nothing interferes, then … (some strict or statistical regularity)'.[9]

Suppose we wish to test $F = ma$ in its unconditional form. We set up a number of different kinds of situations to which, using our bridge principles, we would naturally assign some specific force function. For instance we arrange two bodies of charges q_1, q_2 and masses m_1, m_2 a known distance r apart so that we can assign the force function $Gm_1m_2/r^2 + \varepsilon_0 q_1 q_2/r^2$ directed between them. We ensure as best we can that the situation does not explicitly meet any of the other descriptions to which we know how to assign force functions. Then we also ensure as best we can that there is nothing else about the situation that might be assignable a force function – there is no significant wind, no trucks rumbling by, no bright lights, … Finally we look to see if the motions that occur in all these situations match those predicted from the equation.

Suppose instead that we wish to test 'If nothing interferes with the operation of the force, $F = ma$'. Everything in the description of what we do will be identical to the previous description except for five words. We substitute for the sentence 'Then we also ensure ...' a new one: 'Then we also ensure as best we can that there is nothing else about the situation that might *interfere with the force's operation*'. And what we actually do to ensure this is the same in both cases.

'But,' one may ask, 'how do we know what to eliminate in the second case?' I think the question is more appropriately 'How do we know what to eliminate in the first case?' We do not look for features that figure in our bridge principles as we did in setting up the basic part of the experiment. Of course people who believe in the unconditional form of the law will assume that the features we are looking for are exactly the other things in the situation that can be assigned force functions. But that does not supply them with a method for picking these features out.

Here's what I think happens. We have seen a vast number of cases of forces at work to which we have tried to fit Newton's second law and over time we have established very strong rules of thumb about what can make trouble for it. That is, we have learned from a lot of experience what kinds of things might *interfere* with the operation of the force. That's what we control for.

Earman et al. might think my testing strategy lets in too much. They consider what might seem an analogous reading of *ceteris paribus* laws:

It has also been suggested that we can confirm the hypothesis that CP, all F's are G's if we find an independent, non-ad-hoc way to explain away every apparent counterinstance... But this could hardly be sufficient. Many substances that are safe for human consumption are white; for every substance that is white and is not safe for human consumption, there presumably exists some explanation of its dangerousness ...but none of this constitutes evidence that CP, white substances are safe for human consumption.

The reading that Earman, Roberts, and Smith offer for *ceteris paribus* laws is an excellent attempt at the logical positivist program of substituting acceptable formal-mode concepts for dicey material-mode ones. For example: *X causes Y* becomes *X explains Y*, which in turn becomes *Y can be derived from X given the claims of our theory*. Here we have analogously: *X interferes with $(x)(Fx \rightarrow Gx)$* becomes *X explains why a, which is F, is not G*, which presumably in turn becomes *Ga can be derived from X given the claims of our theory (and perhaps Fa as well)*.

The rendering is a good try, but it does not work, as we can see from Earman et al.'s own example. It does not work for very much the same reason that the analogous formal-mode rendering of causation does not work. You can't get causality and its associated family of concepts out of laws unless the laws themselves are some form of causal laws, not just laws of association, in which case the program of replacement fails anyway.

Moreover, the program is misguided to begin with. There is nothing unacceptable about the concepts of *causation* and *interference*. They are well understood, claims about them are testable and, as G. E.M. Anscombe (1971) argues, some causal relations are directly observable; or, more guardedly, causal concepts do not systematically fare worse in any of these respects than other concepts.

We have, I maintain, ample reason to think that there is as much a fact of the matter about whether it is a causal law that forces cause motions as there is about whether it is a law that $F = ma$; that there is as much a fact of the matter about singular causal claims as there is about other relational claims; and as much a fact of the matter about whether X interferes with some process as about whether the process itself obtains.

The drawback to *interference* is not that there is something wrong with it ontologically; it is rather that we often have epistemic problems. First, we often cannot tell just by looking that X is interfering with Φ, though even this is not always the case. (Sometimes, for example, it is easier to tell that you are interfering with someone's work than to tell that they are working. My friends Anne and Sandy sit at their computers typing just as I do now *for fun*.) Usually to make judgements about interference, we need to have a lot of specialized knowledge and a lot of experience; you can't just tell by looking.

Second, it seems that in most cases there are no systematic rules linking *X interferes with* Φ to other descriptions in some special vocabulary that we prefer epistemically (unless the vocabulary is itself heavily laden with concepts that already imply facts about causality, such as *pushing*, *attracting*, *shielding* . . .). We almost never have 'special interference laws' to tell us in, say the language of masses, charges, distances and times, when something interferes with something else in the way that we have special force laws to tell us when a particular force function obtains. We should note though that the absence of 'special interference laws' is not so epistemically damaging as many suggest. The special force laws do tell us when a particular force function obtains, but only for very specific descriptions – the descriptions that appear in our bridge principles. For other descriptions that may be applied far more immediately and be far more epistemically accessible, such as *a truck passing by* or *the press of the wind*, we are just as much on our own without the help of a system of rules as we are in deciding if we can label the truck passing by as an *interference*.

There are four facts I would like to underline:

1. The lack of systematic rules does not mean that we cannot have knowledge about whether a certain kind of occurrence constitutes an interference. Galileo after all knew to use smooth planes for his rolling-ball experiments because he knew he should eliminate the interference of friction with the pull of the earth. Similarly he knew to drop small compact masses and not feathers from the Leaning Tower. And that was long before he could have had any idea whether friction or the wind exerted a *force* in the technical Newtonian sense.

2. The fact that we cannot identify what counts as interference with respect to a claim Φ does not mean that we cannot test whether Φ is true or not. Consider *Aspirins relieve headaches, if nothing interferes*. We regularly test claims like this in randomized treatment/control experiments.

3. Nor does it mean that is it is too easy to dismiss disconfirmations.[10] When the predicted result fails to transpire, one can always *say* that something interfered. But saying does not make it true. And as epistemologists are always reminding us, saying, even when it is true, does not constitute knowledge, or even reasonable belief. We need a good reason for claiming that something is an interference. When we do not have any idea whether a nominated factor is an interference or not, then we equally have no idea how to classify the case. Our intended test is no test at all.

4. It follows that one needs a great deal of information about what might and might not interfere with a process before we can carry out serious tests on the process and that in turn means that we need already to have a great deal of information about the process itself. That just means that science is difficult, as we already knew, and that it is hard to get started in a vacuum of knowledge.

5. CAPACITIES AND INDUCTION

In *Nature's Capacities and their Measurement* (1989) I offer a number of defenses of capacities:

(a) Once we have rejected Hume's associationist view of concept formation, there is no good argument against the family of concepts connected with causes and capacities.

(b) Strengths of capacities[11] are measurable, just as is the strength of the electromagnetic field vectors or the energy of a system.[12]

(c) We commonly use the analytic method in science. We perform an experiment in 'ideal' conditions, I, to uncover the 'natural' effect E of some quantity, Q. We then suppose that Q will *in some sense* 'tend' or 'try' to produce the same effect in other very different kinds of circumstances. (What I mean by 'in some sense' is that the rules for calculating what happens when a number of factors with different 'natural' effects operate together will differ according to subject matter. Recall the examples of such rules in Section 2.[13]) This procedure is not justified by the regularity law we establish in the experiment, namely 'In I, $Q \rightarrow E$'; rather, to adopt the procedure is to commit oneself to the claim 'Q has the capacity to E'.

In *The Dappled World* (1999) I add another. With the use of capacity language we can provide a criterion for when induction is reliable. This, I maintain, cannot be done with the use of OK properties and strict regularities alone. Earman, Roberts, and Smith object to this claim. This is not surprising because they also reject one of its major premises.

Imagine that we set up a very complex and delicate design, D, an ideal experiment, to observe the precession of a gyroscope in order to test relativistic predictions about the effects of space-time curvature on precession. The result is R. To the extent that we believe our design is a good one and that we have implemented it properly, we expect that that result should be repeatable in just that experimental set-up, i.e., we believe that $D \rightarrow R$ is a strict regularity.

What does it mean that 'our design is a good one'? That is, what criteria must D satisfy if R, which occurs on one occasion of D, is to occur whenever D occurs? The crude answer is that D must control for all factors relevant to R. I read this as 'D is an arrangement in which the capacity of the space-time coupling to produce precession R operates unimpeded'. Those who do not like capacities will try other ways to explain relevance. I imagine they look to strict regularities and I consider two levels at which they might look. First, at a concrete level. Look through all the strict regularities involving very concrete features that have R as a consequent. All and only factors that occur in the antecedents of these are relevant and should be controlled for. My objection to this strategy is not that the list is too long but rather that it will not provide the information we need. Almost *anything* can appear in one of these laws, depending on the arrangement of the other factors; we could design our experiment in indefinitely many ways and still expect the result R. Any feature that was essential to any of these designs gets counted as relevant. Moreover, the long list of regularity laws with R in the consequent will not fix *how* a relevant factor should be controlled for. That will depend on the actual design, D.[14] So lists of strict regularities at a very concrete level cannot provide a criterion that D must satisfy if its results are to be repeatable.

A more plausible proposal is to look at a more abstract level, as Earman, Roberts, and Smith propose. The abstract formula for precession is

$$\text{Precession: } d(n_n^r)/dt = \Gamma^r n_s / \omega_s$$

This formula suggests that an adequate criterion is, 'Eliminate all sources of torque (Γ) except those arising from the space-time coupling as well as all sources of variation in the gyroscope's moment of inertia (I) and in its spin angular velocity (ω)'. Let us concentrate on the torque, as Earman et al., do. They suggest that we couple the formula for precession with 'the laws relating force to precession and various special force laws' to fix what D must be like. What is wrong with that from my point of view? Two things – the first familiar from Section 2. and the second from Section 3.

- The special force laws are not strict regularities among OK features.[15]
- As with Newton's second law, we do not have sufficient evidence to ensure that the precession law can be read as a strict regularity. (A more cautious rendering includes a condition: "If nothing that cannot be described as a torque (or a variation in I or ω) interferes, then. ...")

In their discussion in this volume Earman, Roberts, and Smith deny the first), as we have seen. I suspect they would also deny the second). I have explained why I disagree with them. But if we grant either of these assumptions, we see that the job cannot be done with strict regularities alone. We need capacities. The generalization '$D \rightarrow R$' is reliable because D is a kind of situation in which a stable capacity (the capacity of space-time curvature to affect precession) operates without interference.[16]

NOTES

* Research for this paper was conducted under grants from the Latsis Foundation and the British AHRB, for which I am very grateful. Thanks also to Christoph Schmidt-Petri for help.
[1] I do not mean to imply that I am opposed to them; simply that they are not the kinds of laws I have been thinking about and defending over the last decade.
[2] Or, to be more fair to the proponent of powers, 'whenever the power obtains and the circumstances are propitious for its exercise'.
[3] Note that throughout I take the special force laws to ascribe *forces* but not *motions* to situations.
[4] But they often can make rough predictions or give good advice.
[5] For a more detailed discussion see Cartwright (2000).
[6] For purposes of this section we can remain neutral about my claim in Section 2. It does not matter for the points here whether we take the special force laws to ascribe capacities of a certain types or instead to ascribe an actually existing force.

[7] There are two caveats here. First, it can happen that exactly the same vectorial quantity F that normally is associated with one force law applies to a situation 'by accident' even when it does not satisfy the requisite description because of the particular values of the force properly ascribed to the situation by other special force laws. Second, I would like to remain neutral about how strict we need to be about when 'the description offered in the bridge principle is satisfied'.

[8] Paul Teller (personal communication) has objected to my claim that in quantum mechanics there are only a small number of bridge principles by pointing out that, as I myself urge, there are a good many 'derivative' bridge principles. These, however, almost never expand the scope of the theory, but rather contract it. For they are in fact, as their name says, *derived*. We start with a situation modeled with a combination of descriptions available from our basic bridge principles. Then we add *more* facts about the situation to derive a new force function for it, by limiting the original force function. The derived bridge principle then provides force functions for only a subset of cases that the original did. Of course sometimes we make approximations as we go along. But that, if anything, narrows the scope of the force function even more. For now it is no longer true even of the originally described situation but only of some approximation to it.

[9] In the case of the equations of motion, as we have seen, the caveat really refers to factors not describable in the language of the theory. Setting aside some niceties, we can assume that capacity claims of the form 'A has the capacity to F' imply that if nothing interferes A will F; and probabilistic ascriptions 'A has capacity of strength r to Φ', to imply roughly that if nothing interferes the probability that A will Φ is r. But, as I have argued in Cartwright (1999) capacity ascriptions can say a lot more as well.

[10] For a recent example of this kind of claim specifically in the context of the capacity laws I defend see Winsberg et. al. (2000).

[11] This includes their presence or absence.

[12] We cannot, of course, tell by the measurement itself that what we are measuring is a real capacity, anymore than we can tell by the procedures for measuring the electric field strength that what we are measuring is a real quantity. In both cases that requires a lot of theory.

[13] In Cartwright (1999) I argued that nature might not have always provided such rules. Even in that case there is a cash value to knowledge about the capacity. The associated effects are more likely to occur when a feature with the appropriate capacity is present than when no such feature obtains. (Consider using a magnet to pull a pin from between the floorboards. It is a good idea to try the magnet even should there be no fixed rules for what happens to the pin in just exactly that combination of circumstances.)

[14] In my (1988) and (1989) I give examples where two different laws employ the same factor in different ways.

[15] This is the assumption that figured in the arguments of my (1999), p. 95, where I concluded, "The regularity theorist is thus faced with a dilemma. In low-level highly concrete generalizations, the factors are too intertwined to teach us what will and what will not be relevant in a new design. That job is properly done in physics using more abstract characterizations. The trouble is that once we have climbed up into this abstract level of law, we have no device within a pure regularity account to climb back down again". The device we need includes the special force laws, which, I maintain, can not be rendered as statements of regularity.

[16] In my (1999) I dubbed situations like this 'nomological machines'. This is to highlight both the need to eliminate interference, which I have stressed in this discussion, and the

need to have the right kind of internal structure (one for which there are rules about how the contributions of the parts combine), which I do not discuss here.

REFERENCES

Anscombe, G. E. M.: 1971, *Causality and Determination*, Cambridge University Press, Cambridge.

Cartwright, N.: 1988, 'Capacities and Abstractions', in P. Kitcher and W. Salmon (eds), *Minnesota Studies in the Philosophy of Science, Vol. XIII: Scientific Explanation*, University of Minnesota Press, Minneapolis, MN, pp. 349–355.

Cartwright, N.: 1989, *Nature's Capacities and their Measurement*, Clarendon Press, Oxford.

Cartwright, N.: 1999, *The Dappled World*, Cambridge University Press, Cambridge.

Cartwright, N.: 2000, 'Against the Completability of Science', in M. W. F. Stone and J. Wolff (eds), *The Proper Ambition of Science*, Routledge, London, 209–222.

Earman, J., J. Roberts, and S. Smith: (this issue), ' "*Ceteris Paribus*" Lost', *Erkenntnis*, this issue.

Freud, S.: 1909, 'Notes Upon a Case of Obsessional Neurosis', in P. Rieff (ed.), *Three Case Histories*, Vol. 7 of *The Collected Papers of Sigmund Freud*, Collier Books, New York, 1963.

Glanvill, J.: 1661, *The Vanity of Dogmatizing*, printed by E.C. for Henry Eversden, London.

Hempel, C. G.: 1966, *Philosophy of Natural Science*, Prentice Hall, Englewood-Cliffs, CA.

Hoover, K. D.: 2001, *Causality in Macroeconomics*, Cambridge University Press, Cambridge.

Mach, E.: 1893, *The Science of Mechanics*, Open Court, La Salle, IL.

Ryle, G.: 1949, *The Concept of Mind*, Hutchinson, London.

Winsberg, E., M. Frisch, K. M. Darling, and A. Fine: 2000, 'Review of Cartwright (1999)', *Journal of Philosophy* **97**, 403–408.

Department of Philosophy, Logic and Scientific Method
London School of Economics
Houghton Street
London WC2A 2AE
U.K.

and

Department of Philosophy
University of California, San Diego
9500 Gilman Drive
La Jolla, CA 92093-0119
U.S.A.

MEHMET ELGIN and ELLIOTT SOBER

CARTWRIGHT ON EXPLANATION AND IDEALIZATION

ABSTRACT. Nancy Cartwright (1983, 1999) argues that (1) the fundamental laws of physics are true when and only when appropriate *ceteris paribus* modifiers are attached and that (2) *ceteris paribus* modifiers describe conditions that are almost never satisfied. She concludes that when the fundamental laws of physics are true, they don't apply in the real world, but only in highly idealized counterfactual situations. In this paper, we argue that (1) and (2) together with an assumption about contraposition entail the opposite conclusion – that the fundamental laws of physics *do* apply in the real world. Cartwright extracts from her thesis about the inapplicability of fundamental laws the conclusion that they cannot figure in covering-law explanations. We construct a different argument for a related conclusion – that forward-directed idealized dynamical laws cannot provide covering-law explanations that are causal. This argument is neutral on whether the assumption about contraposition is true. We then discuss Cartwright's simulacrum account of explanation, which seeks to describe how idealized laws can be explanatory.

"One source of misunderstanding is the view … that a hypothesis of the simple form 'every *P* is *Q*' … asserts something about a certain limited class of objects only, namely the class of all *P*'s. This idea involves a confusion of logical and practical considerations: Our interest in the hypothesis may be focused upon its applicability to that particular class of objects, but the hypothesis nevertheless asserts something about, and indeed imposes restrictions upon, *all* objects." (Hempel 1965, p. 18)

In *How the Laws of Physics Lie*, Nancy Cartwright argues that the fundamental laws of physics don't provide true descriptions of how objects behave in the real world. They either make false claims about real objects or true claims that apply only in highly idealized counterfactual situations. For example, when Newton's law of gravitation ($f = Gm_1m_2/r^2$) is interpreted literally, it is false – the net force acting on a pair of objects almost never has the value specified. In defense of this thesis, Cartwright considers two ways that one might propose to reinterpret this law so that it comes out true:

(1)   Interpret the law as including a *ceteris paribus* modifier (i.e., if there are no forces other than gravity at work, then $f = Gm_1m_2/r^2$).

(2)   Interpret the law as describing a component force (i.e., the force due to gravity $f_g = Gm_1m_2/r^2$).

 Erkenntnis **57**: 441–450, 2002.
© 2002 *Kluwer Academic Publishers. Printed in the Netherlands.*

Cartwright rejects interpretation (2) because she denies the reality of component forces; this has elicited criticisms from several philosophers.[1] The correctness of Cartwright's position on this issue is not the focus of our paper. Rather, we want to assess Cartwright's argument that if we interpret the fundamental laws of physics as in (1), then their antecedents will describe conditions that are almost never satisfied; hence, they won't describe how real objects actually behave. With respect to this argument, Cartwright writes:

> *Ceteris paribus* generalizations, read literally without the '*ceteris paribus*' modifier, are false. They are not only false, but held by us to be false; and there is no ground in the covering-law picture for false laws to explain anything. On the other hand, with the modifier the *ceteris paribus* generalizations may be true, but they cover only those few cases where the conditions are right. For most cases, either we have a law that purports to cover, but cannot explain because it is acknowledged to be false, or we have a law that does not cover. Either way, it is bad for the covering-law picture. (Cartwright 1983, 45–46)

Cartwright's argument can be stated as follows:

(3)   The fundamental laws of physics are true only when appropriate *ceteris paribus* modifiers are attached.[2]

(4)   *Ceteris paribus* modifiers describe conditions that hold only under ideal situations.

(5)   When the fundamental laws of physics are true, they apply only to objects in ideal (counterfactual) situations.

(6)   Therefore, the fundamental laws of physics don't apply to objects in the real world.[3]

We will argue that even granting the truth of premises (3) and (4), (5) does not follow (hence, neither does (6)).

Cartwright says that the statement "two bodies exert a force between each other which varies inversely as the square of the distance between them, and varies directly as the product of their masses" is false unless we attach the *ceteris paribus* modifier "there are no forces other than gravitational forces at work". She writes:

> Speaking more carefully, the law of universal gravitational is something like this: If there are no forces other than gravitational forces at work, *then* two bodies exert a force between each other which varies inversely as the square of the distance between them, and varies directly as the product of their masses. I will allow that this law is a true law, or at least one that is held true within a given theory. But it is not a very useful law .... Once the *ceteris paribus* modifier has been attached, the law of gravity is irrelevant to the more complex and interesting cases. (Cartwright 1983, 58)

Thus, according to Cartwright, a true law will have the form '$C \to L$'. Cartwright argues that L is true only if the qualifier concerning C is attached to it, but C is almost never satisfied in the real world. Hence, the law is a true conditional whose antecedent and consequent are both false. For this reason, Cartwright concludes that '$C \to L$' fails to apply to real objects. We now will argue that Cartwright's claims (3) and (4) and a plausible principle concerning contraposition entail that '$C \to L$' *does* apply to real objects.[4]

Let's consider the laws that Cartwright discusses to see if their contrapositives apply to real objects. The first example is the law of gravitation:

> If there are no forces other than gravity at work, then $f = Gm_1m_2/r^2$. ($C \to L$)

This is equivalent to:

> If $f \neq Gm_1m_2/r^2$, then there are forces other than gravity at work. ($\sim L \to \sim C$)

Cartwright claims that '$C \to L$' does not apply to objects in the real world because C and L are each false of real objects. However, this means that $\sim C$ and $\sim L$ are each *true* of real objects, so presumably '$\sim L \to \sim C$' applies to real objects. This leads to the unsatisfactory result that a conditional and its contrapositive, though logically equivalent, nonetheless apply to different things.

The same pattern may be found in another example that Cartwright considers – Snell's law. After claiming that Snell's law is false as it is stated in textbooks, she represents Snell's law as follows:

Refined Snell's Law: For any two media which are optically isotropic, at an interface between dielectrics there is a refracted ray in the second medium, lying in the plane of incidence, making an angle θ_t with the normal, such that: $\sin\theta / \sin\theta_t = n_2/n_1$. (Cartwright 1983, 47)

Cartwright thinks that the condition about optical isotropy is almost never satisfied, so the law does not apply to real objects. However, this means that if a medium is such that $\sin\theta / \sin\theta_t \neq n_2/n_1$, then it is *not* isotropic. Thus, the contrapositive of Cartwright's Refined Snell's Law *does* apply to real objects. If a conditional and its contrapositive apply to precisely the same things, then at least one of these judgments about the applicability of a conditional and its contrapositive must be wrong.

Consider a third example from a different science – the Hardy–Weinberg law in population genetics. It states that:

If no evolutionary forces are at work and the gamete frequency of gene A is p and the gamete frequency of gene a is q (where $p + q = 1$), then the frequencies of the genotypes AA, Aa, and aa are p^2, $2pq$, and q^2, respectively. (Sober 1984)

The antecedent of this law is never satisfied, so if we apply Cartwright's argument, we should conclude that this law does not apply to real populations. However, if we look at the contrapositive of this law, we see that it *does* apply to real populations. If we observe that the genotype frequencies *aren't* at their Hardy–Weinberg values, then there *are* evolutionary forces at work.

Here is the general pattern: take any true law in conditional form whose antecedent and consequent are false (according to Cartwright, all fundamental laws in physics have this feature). In such a case, even if the conditional itself *seems* to be vacuous, its contrapositive won't be, since the antecedent and consequent of the contrapositive will correctly describe real objects. If a conditional and its contrapositive apply to the same things, then either both apply to real objects or neither does.

Thus far we've seen that three claims are in conflict: That '$C \rightarrow L$' fails to apply to real objects if C involves an idealization, that '$\sim L \rightarrow \sim C$' applies to real objects, and that a conditional and its contrapositive must apply to exactly the same things. Which of these claims should be abandoned? To begin with, we think it is implausible to deny that a conditional and its contrapositive apply to exactly the same things. Since a conditional and its contrapositive are logically equivalent, they are different verbal formulations of *the same proposition*. Laws, it should be remembered, are supposed to be extra-linguistic entities; Newton's law of gravitation is no more a part of English than it is of any other natural language. If laws are propositions of a certain type, then Cartwright's position on contraposition must be mistaken.

We also find it implausible to deny that '$\sim L \rightarrow \sim C$' applies to the systems of which $\sim L$ and $\sim C$ are true. After all, scientists *use* such contrapositives to reason about real world systems. For example, if *this population* deviates from Hardy–Weinberg proportions, then *it* must be undergoing some evolutionary process. If this is not an example of "applying the contrapositive to a real object", we don't understand what "applying" means.

What follows, then, is that we should reject Cartwright's thesis that laws of the form '$C \rightarrow L$' fail to apply to real systems just because C involves an idealization. We suspect that Cartwright drew this conclusion by focusing exclusively on the argument form 'If C, then L. C. Therefore L.' Since C is false in the real world, this argument form cannot be applied to real objects. The point about *the argument form* is correct, but nothing follows about *the conditional* itself. For the same conditional also plays a role in a different argument scheme, namely 'If C, then L. $\sim L$. Therefore $\sim C$', and *this* form of argument *does* apply to real objects.

We recognize that Cartwright may want to contest the claim that a conditional and its contrapositive apply to exactly the same things. Given this, it is gratifying that something like Cartwright's conclusion can be defended without taking a stand on this question. Cartwright's main point in advancing her claim about the inapplicability of fundamental laws is to develop a point about explanation: *true fundamental laws do not figure in covering law explanations*. The argument we have in mind concerns forward-directed deterministic dynamical laws – laws that have the form "if C holds at time t, then E holds at $t + \Delta t$". Suppose the C in this law describes idealized circumstances. This means that the forward-directed argument *form* does not apply to real systems, but the backwards- directed argument form *does*, as we have explained. If explanation must be causal and if causes must precede their effects, then the backwards-directed argument, though applicable to real objects, cannot provide an explanation of its conclusion; one can't explain what happens at some earlier time by describing the later state of the system. Recall that Cartwright's goal was to show that fundamental laws don't provide covering law explanations. The conclusion of the argument we have presented is that forwards-directed deterministic dynamical laws that describe idealized circumstances in their antecedents cannot provide covering-law causal explanations, regardless of whether these laws are classified as fundamental or derived.[5]

What is the situation with respect to dynamical laws that are probabilistic? To begin with, we note that whereas a conditional and its contrapositive are logically equivalent, "$\text{Pr}(X \mid Y) = p$" and "$\text{Pr}(\text{not } Y \mid \text{not } X) = p$" are not. Furthermore, it turns out that if "$\text{Pr}(E \text{ holds at } t + \Delta t \mid C \text{ hold at } t) = p$" is a law, then $\text{Pr}(\sim C \text{ holds at } t \mid \sim E \text{ holds at } t + \Delta t) = q$" rarely is. The reason is that laws must be time-translationally invariant; see Sober (1993b) for discussion. The conclusion we draw is that if "$\text{Pr}(E \text{ holds at } t + \Delta t \mid C \text{ hold at } t) = p$" is a law in which C involves an idealization, then the following argument form will not constitute a covering-law explanation:

$$\text{Pr}(E \text{ holds at } t + \Delta t \mid C \text{ hold at } t) = p$$

C holds t

$p \; [================$

E holds at $t \Delta t$

The reason is that the second premise is false. Notice that this point applies, regardless of whether we demand that p be high, as Hempel's (1965) inductive-statistical model requires, or allow p to take any value, which is what Salmon's (1984) account permits.

The conclusion of our argument, then, is that forward-directed dynamical laws fail to provide covering-law causal explanations, if the laws in question are deterministic and contain idealizations in their antecedents, and if they are probabilistic and contain idealizations in their conditioning propositions. This argument differs from Cartwright's in three ways. First, ours does not gainsay the assumption that a conditional and its contrapositive apply to exactly the same things. Second, it does not require a distinction between fundamental and non-fundamental laws. And third, our argument is restricted to *causal* explanations. Despite these differences, we believe that our argument captures much of what Cartwright is after.

Cartwright's goal was to show that laws that contain idealizations cannot be used in covering-law explanations. She thinks that this has implications for many theories of explanation, not just Hempel's, and so she uses the expression "covering-law model of explanation" in a very wide sense, and we have followed her in this. As noted above, the phrase also applies to Salmon's (1984) model. But what, then, does talk of the covering-law model actually cover? The argument we have constructed pertains to any theory of explanation that requires the following: (i) the *explanans* must describe the cause(s) of the *explanandum*; (ii) the *explanans* must cite a law; (iii) all of the *explanans* propositions must be true; (iv) the *explanans* explains the *explanandum* by entailing it or by conferring a probability on it. Forward-directed dynamical laws that contain idealizations in their antecedents (or in their conditioning propositions, if they are probabilistic) cannot figure in explanations, if explanations must have these features. Cartwright (1983) proposes a "simulacrum account of explanation" as an alternative to the covering-law approach; the main point of this account is to make room for the fact that idealized laws can be explanatory. However, she provides very few details on how this new model of explanation is to be understood. We take up this problem in what follows. Our proposal will be that some of the explanations that idealized laws help provide satisfy conditions (i)–(iii), but not (iv).

In evolutionary biology, optimality models describe the value of a trait that maximizes fitness, given a set of constraints. For example, the optimal length of a bear's fur might be modeled as a function of the ambient temperature, the bear's body size, the energetic cost of growing fur, and so on. These models are often interpreted dynamically – if organisms are fitter the closer they are to the specified optimum, and if natural selection is the only force acting on the population, then the optimal trait value will evolve. Understood in this way, optimality models contain idealizations; they describe the evolutionary trajectories of populations that are infinitely large in which reproduction is asexual with offspring always resembling their parents, etc. (Maynard Smith 1978; Sober 1993a).

We want to argue that optimality models are explanatory despite the fact that they contain idealizations. As just noted, these models are interpreted as entailing conditionals of the following form:

(7)   If organisms are fitter the closer they are to the optimal value α and if no forces other than selection are at work in the population, then the population will evolve to a state in which all organisms exhibit the trait value α.[6]

Suppose the optimality model correctly describes how selection acts on the trait of interest:

(8)   Organisms are fitter the closer they are to the optimal value α.

Given this information, suppose we observe that

(9)   The n organisms in the population have trait values $\beta_1, \beta_2, \ldots, \beta_n$ (where each β_i differs only negligibly from α).

Our question is – do (7) and (8), if true, together explain (9)? We think that the answer is *yes*, even if one can provide no details about the nonselective forces that happen to be acting on the population, and no idea how that more complex situation ought to be modeled.[7] Proposition (8) provides a partial description of the initial conditions and proposition (7) provides an idealized model whose antecedent applies to no real world system. Granted, these propositions do not constitute a *complete* explanation of (9) in which *all* causally relevant factors are described, but we think they are explanatory nonetheless.

The pattern here is hardly unique to evolutionary biology. Consider the law of gravitation, understood, as Cartwright says it should be, as describing the net force that would be present if gravitation were the only force at work. If the law plus the true masses of a pair of objects and the distance between them and the assumption that no other forces are at work (plus $f = ma$) entail that the objects should exhibit an acceleration of α and one observes that the acceleration is β (where α and β differ only negligibly), then the idealized law plus the partially specified initial conditions are explanatory.[8]

If (7) and (8) do explain (9), the idea that explanations are arguments appears even more doubtful than Salmon (1984) argued that it is. Salmon's point is that the *explanans* can confer a low probability on the *explanandum*. However, we don't think that (7) and (8) confer a probability on (9) at all. What is the probability that the observed trait values (the β_i's) will be close to α, given that α is the trait value that should evolve in an idealized circumstance that does not obtain? We don't know, but it isn't necessary to know this. Propositions (7) and (8) explain (9) even though they do not tell you what the probability of that proposition is. In this type of explanation-by-idealization, conditions (i)–(iii) are satisfied, but (iv) is not.

We began by criticizing Cartwright for drawing an invidious distinction between a conditional and its contrapositive. We then showed how her argument can be reconstructed without requiring that a conditional and its contrapositive apply to different things. This new argument reaches a slightly different conclusion from Cartwright's; we showed how certain sorts of dynamical laws cannot figure in covering-law explanations that are causal. We then tried to flesh out Cartwright's idea that explanation-by-idealization requires a new account of explanation. A causal model contains an idealization when it correctly describes some of the causal factors at work, but falsely assumes that other factors that affect the outcome are absent. The idealizations in a causal model are *harmless* if correcting them wouldn't make much difference in the predicted value of the effect variable. Harmless idealizations can be explanatory, as is shown by the fact that (7) and (8) help explain (9). In this pattern of explanation, the *explanans* is entirely true; it explains the *explanandum*, not by entailing it or by conferring a probability on it (high or low), but by showing that the value described in the *explanandum* is close to the value predicted by the idealization.

ACKNOWLEDGEMENTS

We thank Michael Byrd, Nancy Cartwright, John Earman, Ellery Eells, Berent Enç, Branden Fitelson, Malcolm Forster, Daniel Hausman, and Marc Lange for comments and suggestions.

NOTES

[1] For this issue see Forster (1988a, b), Creary (1981), Chalmers (1993), Earman and Roberts (1999), and Needham (1991). Although all these papers are somewhat relevant to Cartwright's argument above, Forster and Creary specifically address Cartwright's challenge concerning the reality of component forces.

[2] Although Cartwright says that all laws require *ceteris paribus* modifiers, it is clear from her discussion that in the case of the fundamental laws of physics, she thinks the *ceteris paribus* modifiers are exactly specifiable. This contrasts with the views of philosophers (e.g., Schiffer (1991)) who think that the *ceteris paribus* clauses used in special science generalizations aren't exactly specifiable.

[3] Cartwright (1999) says that she still holds the view of laws she defended in Cartwright (1983).

[4] It is arguable that contraposition is not always valid. Consider the following example: "If I made a mistake, then I didn't make a big mistake". The contrapositive of this conditional is "If I made a big mistake, then I didn't make a mistake". However, contraposition is valid for the laws we will consider. Detailed discussion of this issue can be found in Jackson (1991).

[5] What if a *backwards* deterministic law contained an idealization in its antecedent? The law will have the form "if I holds at time t, then C holds at time $(t - \Delta t)$", where I involves an idealization. The contrapositive of the law says "if not-C holds at $(t - \Delta t)$, then not-I holds at t". If not-C and not-I both apply to real systems, then this law can be used in a Hempelian explanation. And if one does not prohibit "negative properties" such as not-C from being causes, the contrapositive seems capable of providing a covering-law *causal* explanation. This is why, in the case of deterministic laws, we have limited our argument to *forward*-directed laws that contain idealizations in their antecedents. We owe this point to John Earman.

[6] Talk of all other evolutionary forces being absent sometimes means that some quantitative variable has a value of zero (e.g., the mutation rate), but at other times it means that certain idealizations are in place (as in the assumption of asexual reproduction). As Cartwright (1983, p. 45) says, it is sometimes apt "...to read '*ceteris paribus*' as 'other things being *right*'".

[7] This answer does not involve a commitment to adaptationism, which can be thought of here as the view that (7) and (8) are usually true if (9) is; for discussion, see Sober (1993a). Nor does it oblige one to accept the following generalization:

> If organisms are fitter the closer they are to the optimal value α and if the forces other than selection are of only negligible value, then the organisms in the population should exhibit trait values close to α.

It is possible that the true but unknown underlying laws exhibit sensitivity to initial conditions.

[8] Selection and Newtonian gravitation are each construed as deterministic forces within their respective theories. When an idealization concerns a force whose effects are described probabilistically, α should be interpreted as an expected value.

REFERENCES

Cartwright, N.: 1983, *How the Laws of Physics Lie*, Clarendon Press, Oxford.

Cartwright, N.: 1999, *The Dappled World: A Study of the Boundaries of Science*, Cambridge University Press, Cambridge.

Chalmers, A.: 1993, 'So the Laws of Physics Needn't Lie', *Australasian Journal of Philosophy* **71**, 196–207.

Creary, L. G.: 1981, 'Causal Explanation and the Reality of Component Forces', *Pacific Philosophical Quarterly* **62**, 148–157.

Earman, J. and Roberts, J.: 1999, '*Ceteris paribus*, There are No Provisos', *Synthese* **118**, 439–478.

Forster, M.: 1988a, 'The Confirmation of Common Component Causes', in A. Fine and J. Leplin (eds), *PSA*, Vol. 1, pp. 3–9.

Forster, M.: 1988b, 'Unification, Explanation, and the Composition of Causes in Newtonian Mechanics', *Studies in History and Philosophy of Science* **19**, 55–101.

Hempel, C.: 1965, 'Studies in the Logic of Confirmation', in *Aspects of Scientific Explanation*, Free Press, New York.

Jackson, F. (ed.): 1991, *Conditionals*, Oxford University Press, Oxford.

Maynard Smith, J.: 1978, 'Optimization Theory in Evolution', *Journal of Ecology and Systematics* **9**, 31–56. Reprinted in E. Sober (ed.), *Conceptual Issues in Evolutionary Biology*, MIT Press, Cambridge, MA, 1994.

Needham, P.: 1991, 'Duhem and Cartwright on the Truth of Laws', *Synthese* **89**, 89–109.

Salmon, W.: 1984, *Scientific Explanation and the Causal Structure of the World*, Princeton University Press, Princeton, NJ.

Schiffer, S.: 1991, '*Ceteris paribus* Laws', *Mind* **100**, 1–17.

Sober, E.: 1984, *The Nature of Selection*, The University of Chicago Press, Chicago and London.

Sober, E.: 1993a, *Philosophy of Biology*, Westview Press, Boulder, CO.

Sober, E.: 1993b, 'Temporally Oriented Laws', *Synthese* **94**, 171–189. Reprinted in E. Sober, 1994, *From a Biological Point of View*, Cambridge University Press, New York.

University of Wisconsin-Madison
Department of Philosophy
5185 Helen C. White Hall
600 North Park Street
Madison, WI 53706
U.S.A.
E-mails: melgin@students.wisc.edu
         ersober@facstaff.wisc.edu